LIZ PEARSON MANN

Eat Like Your Ancestors: From the Ground Beneath Your Feet

A Sustainable Food Journey Around the English West Midlands

First edition

ISBN: 978-1-8381830-2-8

Editing by Anna Bliss
Editing by Andrew Dawson
Cover art by Nikki Ellis
Illustration by Laura Templeton

This book was professionally typeset on Reedsy.
Find out more at reedsy.com

To Andy and family
(former inhabitants of Red Rock Farm)

Contents

Preface

Food is important, and on that I'm sure we'd all agree. It's at the heart of cultures the world over because it's so important in our lives.

We spend a lot of time thinking about, shopping for, making and talking about food. You may have bookshelves groaning under the weight of cookbooks, and you might watch hours of food programmes on TV, so I'll bet you're not short of images of delicious food, recipes and inspiration.

Despite all this information though, and the beautiful cookbooks, if you're reading this, you're probably also worried about food. You keep hearing about our massive food footprint and it's likely that you, like most of us, face two opposite problems. Either too much abstract information, or overly simplified calls to action.

Much of what you've been reading concerning sustainable food speaks of statistics, statistics, and yet more statistics. Scientific data and 'facts'. Graphs with cow and plant symbols and kilograms of methane per kilogram of food. Invisible gases which only the experts understand.

At the other end of the spectrum, you may see food reduced to a single tick box at the end of a long list of 'to dos' for living a green and sustainable life. We used to see the advice 'Eat Local and Seasonal'. The advice hasn't entirely gone away, but now the words 'Eat Less Meat' have risen to the top of the pile for a quick takeaway on sustainable eating. Life is supposedly simple.

But I've been wondering, where are the words about the farmers who grow our food? Of their farms, and the landscape in which they produce that food? Of animals, pasture and cornfield – the stuff of our childhood learn-to-read books? We can find them, but other voices speak louder.

To top it all, everything you read might seem dismal, as if eating all the good food you like must come to an end. But I feel this viewpoint arises from a mindset that's stuck in the worst of the present. Quite the opposite of being stuck in the past. Yet far from it. The past has much to show us about the potential future of food.

The worst of the present is the industrialised farming of animals and crops alike. It's the reporting on this worst present which has pushed out of our minds ways of producing food which have a long history. Many of those ways continue to this day. You might just be unaware of them because popular opinion has pushed this knowledge out into the sidelines.

This book is not all about statistics. You might breathe a sigh of relief!

What This Book is About

This book is for green-living enthusiasts, like me, who need to hear a different story. The tired of statistics, the curious: those who want more than a single tick box. And also the nature enthusiast (or the would-be nature enthusiast), because nature is an important part of the story.

This is a journey around the English West Midlands where I live, through distinctive landscapes that produce distinctive food. Food from the small farm on which my husband grew up features, as do some of his family reminiscences I've heard over the years.

Even if you don't live anywhere near here, this can be your journey too, as much of what we come across should ring bells with you, wherever you are. And the takeaways you can use in your kitchen too.

This is the antithesis of the planetary diet of which we hear so much now. You may have heard of this. It's a diet proposed in a report by the Eat-Lancet Commission in 2019. The authors of the report say that we must drastically reduce the amount of meat, particularly red meat (for instance, beef, lamb or venison), dairy, eggs and starchy vegetables that we eat. Their aim is to protect our health and the environment.

This report promotes a diet which you would eat no matter where you live. You could live in the desert, the tundra, a wet Welsh hillside or a hot, dry Mediterranean island; you would all eat a similar amount of each food group, and they guide your food by a reference diet which details the exact number of grams per day of each food type.

This book doesn't give you one diet to follow, measured exactly down to grams of one food group or another. It is, instead, about the foodscape outside your front door, which may be quite unlike the foodscape outside many other people's front door. Understanding it is the foundation of sustainable eating, and yet, eating that doesn't feel like following school rules.

It's also a story of ancient food that lies under the ground. Archaeologists unearth food remains from the ground every day around me, where developers build supermarkets, housing estates and road-widening schemes at a pace.

This is stuff that I deal with every day at work as an archaeologist, working in commercial archaeology. Here we have the principle that the polluter pays, so should planners say so, developers have to bring in people like us to dig up remains under the ground before they start their development. They pay, we dig. We're unearthing the rubbish left behind by people like you and me through the ages: their food waste, their cookware, tableware, and farm waste. Their farming ways, too, are etched into the ground.

The same may happen where you live. Whether archaeologists in hi-vis jackets and hard hats descend upon those sites regularly, though, depends on the country in which you live.

Talk to anyone on our team (particularly the hard-hat and hi-vis wearing bunch) and if we've stayed in one place for long enough, we gain something akin to a bird's-eye view of all the activities going on over time. It's a bird's-eye view of farmsteads appearing, disappearing, and moving around in the landscape; of the pattern of cornfield, woodland, pasture, haymeadows; of drovers walking animals along droveways across ridgeways to distant farms and cattle markets; of barns, dovecotes and windmills; of orchards and hop yards.

It's all part of the story, and it shows us why our local food cultures are what they are. For they've often developed over hundreds, and even thousands of years in sympathy with the local landscape. There's wisdom in those cultures. Wisdom that well-meaning individuals, corporations, researchers and think tanks encourage us to reject.

Over the past few years or more, it seems as if we've been in a frenzy over the question of what makes sustainable food, and over time I've realised that perspectives, information, and words from my 9-5 work have a bearing on all this. This is where it started; when I thought about writing a book because I felt we need another angle on what is sustainable food. One that we see from the ground up, and with time perspective too.

It's not what I thought I would do - take my work home with me and into blog and social media posts. I was writing about frugal ways with old wool (for knitters), or plastic-free shopping, 1970s style: a trip down memory lane, shopping with my nan. And, so this book was born.

These are my perspectives based on a landscape that I know well. I'm not a farmer (few of us are), but I've spent many years in a peculiar job, producing data on the lives of farmers and the food they've produced from the ground beneath their feet. And it is that ground beneath our feet from which we've become very disconnected. It's no surprise - seeing as many of us now live in cities and on suburban housing estates. Most of us have never spent our days producing food.

* * *

Few of us have grown up in the country among farming families. I didn't either. I grew up on a 1970s housing estate on the edge of town. I lived on the edge: the edge of town and country, and of two different food-producing landscapes. Behind my house was more 1970s housing estate, but at the top of the road there was a gate that led into fields of corn.

I remember little about those fields of corn, apart from a brief foray into them with my friend and her dog, for which we were ticked off for trespassing and flattening tracks through the corn. One of our mums had to apologise to the farmer, whilst we hung our heads and pretended the dog took us there. Now, more housing estate, and further away, a university science park have encroached upon these fields. This was not my playground, although later my attention turned very much to crops.

My playground was meant to be the housing estate where I lived, but I had a tendency to sneak off in a different direction from the tarmacked roads and flat cornfields, and across a main road where the land rose up into the forest.

Uphill I would go through a craggy landscape of woodland and small paddocks where sheep were grazing. Cycling uphill on a shopper-style bike that only had three-speed Sturmey-Archer gears was my challenge. It wasn't the best bike for this, but I would grit my teeth, get so far then collapse over the handlebars to catch my breath. There would nearly always be a couple of sheep perched atop a crag, munching grass and looking down on me with disdain. I would get back on my bike and keep going to where the land plateaued. Cycling along, buffeted by the wind, I was happy enough, surrounded by a landscape of sheep, cows and a few fields growing cereals. It's still like that today.

On the way back downhill, I would dive into the woods for a wander. This was woodland with public access, so at least I wasn't trespassing here. Back at home, my mum would interrogate me, asking 'Where have you been?' I would mumble something like 'Oh, not far. Only just up the road.' But, I'm sure the bits of bracken falling out of my sleeve cuffs, and the acorns and pine cones in my pockets from my woodland visits were a real giveaway. I was meant to stay closer to home but I preferred the steep climb, the windswept plateau, and manure smells, to the flat wheat and barley fields.

This was further east than I live now, on the edge of Charnwood Forest in Leicestershire. I doubt I thought much about my forest playground as a food-producing landscape, or understood why my house stood at the junction of two justifiably different farmscapes. My house stood on flat fertile soil that can produce grain well, but in front of the house was a great slab of volcanic rock, with craggy edges, that formed the 'forest'.

Going back into the past, because of the terrain and not-so-fertile soils, it stayed a wild landscape. It was more hunting ground than farmland for much longer than anywhere else. Its past and present is of deer park and pasture: the wool from sheep financing the cottage industry of weavers, whose cottages are still in the villages today.

If you live around here, you can get your free-range milk and meat from local farmshops, or visit an alpaca farm.

Maybe you have memories like this too. You had time, you could wander and observe, but for many of us, unless we go to work on the land, we find ourselves drawn away. Our time is taken up with getting to work, working to earn a living, and all the myriad of responsibilities that take up that time. It's easy to lose touch. You find that food (in your mind) doesn't come from the land or the sea. It comes from the supermarket, which we whizz around, dodging other people's shopping trolleys, navigating the checkout. We have detached ourselves.

* * *

This is what this book is about: ways to reconnect. At the end of each chapter are takeaways which you can put into action at home. They will change your mindset. For changing your mindset is at the at the heart of making change. Reject black and white thinking. There is no one magic trick or hack that will help us solve all our food-related problems.

You might wonder, is this just chocolate-box England? A rose-tinted past to which we cannot return? Or are our modern ways leading us to a point of no return? I can't pretend to know all the answers, but they're questions worth asking.

Above all, you may ask, WHAT ABOUT THE METHANE? Read **Chapter Six** for a dig beneath the surface of the most influential statistics circulating in the media to get an idea of why all is not as it might seem.

We travel round my local landscape in a roughly clockwise direction, through the seasons, coming across different farm-scapes and different challenges for producing food. Hence, different food cultures.

We start by going **Into the Hills** in **Chapter One**, in south Shropshire on a cold, chilly January day. We find out why much of this landscape has always remained under permanent pasture, and we discover the food at the heart of farming there.

At Easter, in **Chapter Two,** we move round and stay in the hills. We visit **Red Rock Farm**: a small farm where my husband grew up. We're still with the animals, and we hear much about why those red rocks and the lie of the land have shaped the farming and food produce there for aeons. Hear why you could buy from and support hill farmers who are custodians of rare breeds and some of the most wildlife-friendly countryside of the British Isles.

In **Chapter Three**, we move round onto flatter land: land of **Cow and Corn**. We go back in time. We find ancient food under the ground. We eavesdrop on people drying their grain ready for storage. They work with the principle that diversity wins in a world of uncertain weather, and where it's a challenge to keep up soil fertility in an area where poor soils are common. Adapt to rustic, but diverse, food that is closer to a long-lived pattern of eating.

The weather is warming up. Spring is here, and in **Chapter Four**, we're in the cornfields; **The Breadbasket** of the West Midlands, moving through a landscape of windmills, barns and corn dryers. We find out why this is a small area of hard-pressed land, and why ancient and heirloom grains are creeping back into fields and into our kitchens.

It's high summer, cooling into autumn. In **Chapter Five, Terroir: Vegetables, Cider and Beer** we visit an area where small market garden smallholdings and orchards, now morphed into larger market garden farms, used to dominate. We come across an area famed for its diverse, artisan food and drink; for local varieties and the 'terroir'. All this talk of 'better' food. It begs the question, is better food affordable?

At the end of each chapter you'll discover ways to reconnect. These are ways that I realise I've gone through over recent years. The **final** chapter brings these ways together, to help you decide what to do in a way that feels right for you and where you live. These are just prompts to set your mind working. In the end, we all need to adapt any advice we see to our own situations. You can make your roadmap to eating well without trashing the planet, but still respecting the lives of animals (farm animals and wild animals).

You will also find recipes, with tips on introducing more sustainable ingredients into your food, which you can do gradually. Before long, you will find this is an automatic thought pattern.

This is a short book because I doubt that you want to read the equivalent of a PhD in gaseous chemistry, or that you're particularly attracted to a greenhouse gas emission diet. To feel rooted and grounded in the food belonging to the landscape and culture in which you live. This is a mindset in which to immerse yourself in.

Acknowledgement

Thank you to all those who have helped me put this book together.

To editor Anna Bliss for her assessment of my first draft, and insightful comments, which led me forwards to the final book. I'm grateful for her enthusiasm for my wanting to lead you, the reader, to take delight and interest in food, the landscape from which it came, farmers and smallholders, and all people who produce food. For it is not enough to simply think about food in terms of a 'carbon footprint', if you want to eat well but tread lightly on the planet. Facts and figures can distort our perception of real life, and sustainability alone is unlikely to make us commit.

To Andrew Dawson for proofreading, shaping and fine-tuning the text.

To Nikki Ellis for working with my initial ideas and coming up with the cover design which conveys the meaning of this book well.

To Laura Templeton for putting together the map of the sustainable food tour at the beginning of this book. She knows the area well and has put together a map with a feeling of place and character.

All have weathered, with friendliness and support, my first foray into Indie Publishing, which has been made, very much, with my L-plates on.

Thanks also to the Mann family. Firstly to my husband, Andy, for putting up with hours of my typing away at this book, and endlessly scratching my head over each stage of publishing. I've asked him many questions about Red Rock Farm, usually at inopportune moments, such as when he's engrossed in the final episode of a series of *Star Trek*. Questions such as "So, going back to the farm - what did you feed the pigs on?" are very at odds with a world operating at warp speed on TV. Thanks also to the Mann family for their reminiscences of escapee sheep, farm wildlife, and questions from me about rock house-dwelling relatives.

I also owe much of this book to years of working in archaeology. Looking at long-buried food waste day after day can feel very remote from everyday life, but it comes to life when I think of those food-producing ways of old, and farming ways etched into the landscape, that have much to show us today. It's very relevant to the debate going on about how we are to feed ourselves into the future. I spend much of my time indoors, but my colleagues spend their days in the mud, wrestling evidence out of the ground that I can portray in this book. So it's on the back of their toil that I wrote some of the content you see here.

I thank Suzi Richer and Aisling Nash for photos of sheep and furrows in the land, respectively.

Thank you also to the following publishers, organisations and authors who have allowed me to quote or include their work: Hayloft Publishing, Elliot and Thompson, Permanent Publications, Worcestershire Archive and Archaeology Services, Wade Muggleton and Clare Tibbits.

Telford and Wrekin

SHREWSBURY

River Severn

Shropshire

The Wrekin

CHURCH STRETTON

The Long Mynd

Wenlock Edge

Brown Clee

LUDLOW

Clee Hill

KNIGHTON

River Teme

LEOMINSTER

Herefordshire

River Wye

HEREFORD

Malvern Hills

The Peak District

STOKE-ON-TRENT

River Trent

Staffordshire

STAFFORD

LICHFIELD

River Tame

KINVER

Birmingham and the Black Country

RED ROCK FARM

West Midlands

Worcestershire

AMBRIDGE

STRATFORD-UPON-AVON

WORCESTER

PERSHORE

Warwickshire

BADSEY

Vale of Evesham

Bredon Hill

TEWKESBURY

The Cotswolds

XV

1

Into the Hills

The hot topic, at the moment, for living a green and zero waste lifestyle revolves around what to eat, and what makes a sustainable diet.

All the focus is on whether you eat meat, how much of it, and whether dairy is ok, wherever you look. You might be reading this because you want to know what to eat. Eat meat/don't eat meat/become vegetarian or pescetarian? I wonder, though, have we become too tied up in knots by one single question?

Instead of starting here, I've been thinking more about the ways of homesteaders and farmers. About their connection to their environment too, for it is the solid rock beneath their feet, the water lapping up against their shores or banks, the soil and the terrain, that influences what they produce. And, what makes a sustainable farmscape and diet. It is rock and soil, though, that I focus on in this book, as I travel around an inland landscape.

A Pastoral Landscape in January

Bearing this in mind, I bring you to a day in early January when I happened to be in Shropshire, one county north of home. I was in a small town called Church Stretton. I stood in the middle of town on a chilly, windy day looking at the surrounding hills. The high ridges of the Long Mynd and Wenlock Edge were in the near distance, Brown Clee hill beyond, and the faint baa of sheep all around. Cows were grazing on lower ground, and there was an occasional field showing the remains of crop stubble, but not many. It was a classic pastoral scene.

Just after the Christmas and New Year binge, I was feeling the effects of too much rich Christmas food and had been thinking about bringing lighter dishes back on to the menu.

More vegetables, fruit, nuts and grain wouldn't go amiss in our household. After all, my husband and I have an allotment. It occurred to me that if we could spend more time growing some of our own food, we'd have low-cost, nutritious produce coming straight into our kitchen. We get grand ideas about what to do on the allotment; some of which stick and some of which fall by the wayside, so we're constantly revising our approach. But either way, my thoughts were on how to maximise our use of this food.

Recipe Ideas: Ingredients, Seasons and Food Miles

I spent some time perusing vegetarian recipes, and thought about the ingredients most common in those recipes. They nearly all included avocados, quinoa, chickpeas, kidney beans, sweet peppers, and almond milk. Scanning down that list, at the time it sounded ok. Apart from the almond milk—that's never been quite my cup of tea.

One issue, though, was that one of my New Year's resolutions was to eat more locally and seasonally, and it didn't fit. I'm not religious about it, and I speak not from a pedestal of perfection. After all, I like a bowl of guacamole as much as the next person, but I don't want to be heavily dependent on imported food.

It occurred to me that I ought to do some food shopping. I peered into a greengrocer's windows. I would have gone in and nosed around, but they were closed as it was a Sunday morning. I hovered outside the local convenience store, a Co-Op, wondering what to buy. I could have bought all those ingredients listed in those recipes in the convenience store, but something made me move on.

You may live somewhere warmer and sunnier, with a good range of fruit and veg available all year round. None of those ingredients, though, grow well (or even at all) in rainy old England. They weren't making it onto the plate at home during winter, and I didn't want to start that day.

In our household, we've stuck to mostly local and seasonal, and fairly plain fare. Behind it all, our default diet has basically been of the 'meat and two veg' variety as that's what we've both been brought up on.

Standing on the main street of Church Stretton, that traditional British diet fitted the scene. It was winter, there was a chill wind, and sheep were bleeting on the hills. We hear so much, though, about the travesties of how some animals are farmed today. We see images of pigs confined in crates, barely able to move, and we hear about the amount of cropland used to feed those animals, so you might question the bucolic scene.

The world's attention, at the moment, is on the land used to grow soya in the Amazonian rainforest (particularly in Brazil), to feed livestock. What's rarely mentioned is that most of the soya used in animal feed is made from waste left over after processing to make soya oil and other products. Considerable quantities of soya feed do come into the UK to feed livestock - mostly for poultry and pigs, rather than our grass grazers (mainly sheep and cattle).

But, those are modern farming ways, and here I stood in traditional livestock farming country. It had me wondering, if our parents and our grandparents lived on meat, dairy, grain and vegetables, were they wrong? And what of the generations stretching out before them? Considering current opinion, you might wonder what on earth they were doing. How could they have got it so wrong?

We could do with asking them. What would they say - if we said they were wrong? We need to go back in time, and back to the land. After all, the further we go back, the more our ancestors lived from the ground beneath their feet. Wherever you live, your food comes from a landscape, so I'm thinking about this from the ground up. We're in pastoral countryside, so we could take a look around where I live, the West Midlands.

Iconic Pastoral Landscapes of the British Isles

Apparently, livestock farming uses up extensive land, but here I was looking out on extensive, hilly pastoral land in Shropshire. We have plenty of it in the British Isles. Think of the Pennines, the high Lakeland fells, the Brecon Beacons, Snowdonia, and the Cairngorms.

There is plenty of hilly pastoral land elsewhere in the world, and maybe where you live too. According to the World Atlas, 24% of the earth's landmass is mountainous. And that's before we've even accounted for all the land simply described as hilly or 'upland'.

Ryeland sheep; a breed common to the borderlands of Herefordshire, Shropshire and Wales

These hilly or mountainous areas of the British Isles are iconic, but they're not the landscapes that stock supermarket shelves with tomatoes and salad leaves.

Can you imagine this landscape turned over to waving fields of corn, soya and market garden vegetables? Even if we think it's more efficient to turn land over to crop production, the uplands are hardly the place to do this on a large scale.

If we go up into the hills here, winters are long and cold, and the growing season is short. If you were an upland farmer, there's a good chance that you'd be dealing with poor soils, so it would be hard for you to keep producing crops season after season. It's no small wonder that livestock are still the mainstay in the hills.

Even so, upland farmers and smallholders have, in the past, had to produce some crops for self-sufficiency. Smallholding, by the way, means a piece of land smaller than a farm (less than 50 acres) but larger than an allotment. It's only in recent years that food production has become more specialised, with more separation between animals, corn, fruit and vegetables.

Now we're hearing the call for farmers to return to mixed farming, and when it comes to smallholders, it's always been so. It's all about balance. Not a 50:50 balance between crops and livestock, but a balance that works with the landscape, and crops that suit the terrain and climate. There's an oaty flavour to the countryside around here, as oats live through cold, wet weather in the hills.

Herdwick sheep in the Lakeland fells. Courtesy of Suzi Richer
@EnviroSuzi on Twitter

Some upland farmers and smallholders are adopting forest garden techniques to increase the range of crops they can grow.

Like Katie Shepherd[1], who writes for the UK Permaculture Association and has adopted permaculture techniques in the hills.

She's brought traditional hardy livestock back onto her land, which are more contented living outdoors throughout most of the year, and will subsist on rough pasture and hay from her haymeadow. She's also started a small-scale forest garden, where manure boosts nutrients in the soil and keeps them warmer in winter. Shelters and windbreaks create a better environment for vegetables and fruit trees than would an exposed hillside. This is small-scale plant food production in an otherwise pastoral landscape.

If it made sense to those working with the natural landscape to concentrate on farming crops, then that would be our historic landscape up in the hills. That it isn't so should be telling us something.

Good Food, Wild Country

We were looking out on sheep feeding on the surrounding grassy slopes. They were not feeding on feed cake made of grain and soya that is factored into most statistics on emissions related to meat production. Somewhere else, they may be; but as far as I could tell, not here.

If I'd have walked to the edge of town and been able to find a farmer, I could have asked "Excuse me, do you feed your sheep on soya from the Brazilian Amazon?" I'm sure they would have

---

[1] *Hill Farming – A Permaculture Perspective* by Katie Shepherd, in Permaculture magazine; https://www.permaculture.co.uk/articles/hill-farming-permaculture-perspective

given me a funny look. What a strange question to ask. But then, with all the focus on soya from the Amazon being used in animal feed, you might think that all the world's livestock are fed on Amazonian soya.

Here sheep were on hillsides, eating grass and wild herbs. I'm assuming this isn't just a pastiche. A display for visitors during the day, after which they're brought in doors and fed on industrially-produced livestock feed. It's highly unlikely. Further south in Chapter 2, I had a farmer to ask, and you might be surprised at the answer. Read on....

It was not only grass they were feeding on. In this landscape, there are many types of grass, sedges, spike-rushes in boggy hollows, and hawthorn and blackthorn leaves in the hedgerows. Birds of prey, such as red kite, fly above and animals rustle in the undergrowth. There's wildlife up in those hills. More so that you would find in the average modern arable field.

Sheep droppings and cow pats hit the ground, and immediately the dung flies descend. As long as certain parasitic wormers haven't been used, the dung flies burrow and tunnel down, carrying organic matter and nutrients into the soil as they go. Dung fungi follow on later, finishing off the process. The soil is alive with insect critters, worms, fungal hyphae, and more.

Why not take a good look? You'd see small mammals snaffling, and birds pecking at seeds and insects on the surface. Stay long enough, and you'll see larger animals; look above, and birds of prey will pounce and swoop on smaller animals.

Walk into a modern arable field, and you'll find a quiet field. There's much going on that you can't see. Fed with chemical fertilisers, not poop, soils are losing organic matter and nutrients. Crops are doused with herbicides and pesticides multiple times before harvest. The soil is virtually inert, yet it's hard for us to see.

Hide out on pasture up on a hilltop (or on pasture anywhere) and during a rainstorm there's a lot going on that's also hard to see. Organic matter in the soil, and a dense mat of roots, will be sucking up water. Your patch of pastureland will be becoming like one big green sponge. All the better for you if you live anywhere that's prone to flooding. Especially if you live near a major river.

I live close to the River Seven - the longest river in England, in a place that can flood badly some years in the winter. Starting in the Welsh hills, flowing past Shrewsbury, Worcester, Tewkesbury and out into the Severn Estuary alongside Bristol, it cuts through a long floodplain. I'd rather farmers upstream kept their pasture (their big green sponge) intact. But I'm hearing that farmers are ploughing up their pastures. Who can blame them?

We, the unwitting general public, have bought into the idea that red meat is bad for our health, and that cows are heating up the planet with their methane. We've been hearing a steady stream of news connecting meat and ill-health over decades, but the news is coming in that much of this research is unreliable. Meat from animals that have been wholly or mostly grass-fed have a much healthier balance of fats, and we now know that fat doesn't make you fat. In fact, the finger is now pointing at sugar and refined flours.

We're scared about the connection of meat with cancer too, but now we're hearing that modern ways of preserving meat with chemicals are more to blame. Yet, for hundreds and thousands of years people have been preserving meat and fish using plain old simple resources: air (drying), a smouldering fire (smoking) and salt from the ground or sea. Yet, there's no evidence that those processes cause any problems, but it takes time for that news to filter through to us.

You might want to watch your burning of meat on the barbeque, though, as I gather that too much burning degrades fat in meat in a way that our bodies don't tolerate well. I love a little singed sausage from the barbeque, but I'd best keep that as a summer treat.

With all this bad publicity surrounding meat (particularly red meat), what's a livestock farmer to do?

A growing band of farmers are rising to the challenge. They're now turning back the clock and going back to feeding their animals on pasture and hay, and on 'unimproved' pasture at that. They're not seeding their pasture with a monoculture of a single grass species (usually annual ryegrass), so the natural herb-rich grassland can regenerate. They're shunning chemical fertilisers too, that let a few grass species grow at the expense of a diverse grass-herb community.

They're selling grass-fed meat direct to people like you and me. They have websites that explain why their wildlife-rich pastures keep the animals in good health, and us too.

We can play our part. There's always something we can do to influence change for the better, in a way that fits with our circumstances. If they've reinstated hedgerows, and barn owls are roosting in their barn again, they'll tell you so. More farmers are looking to make these changes.

What Can We do?

We can:

Check Out Our Food Miles

We need to eat locally, as not only does it squash down our carbon footprint, it's a reviver of local farming communities and their skills. Among their community are those making cheese, potted or smoked meat and other delights. We might think we eat local, but we could be blind to the amount of imported industrially-processed food that we really eat.

If you like precision and hard data, you could try the Food Miles Calculator[2] in the Notes section. Otherwise, you could just eyeball your shopping lists, and do some spot checks on where that food typically comes from, to get an overall idea of what your food footprint really is.

Once you've gone more local in the kitchen, take more note of the imported food you still want to buy. Most of us would feel a little deprived if imported food or drink never passed our threshold, as for centuries we've been importing some of what comes into our kitchens. But the world has limits. We could think more of this being variety that is the spice of life.

After all, Romans brought quite a few exotics into our country, and some of this lifestyle stuck. We like wine, and we grow some, although the right micro-climate for vineyards in the British Isles is limited today. We may have lost the habit of drizzling fish sauce (fermented fish and anchovies) on our dinners, but we import other exotic sauces instead. Dormice are not a thing on

---

[2] Food Miles Calculator; https://www.foodmiles.com

our dinner tables. Elite Roman dishes served up on villa estates have faded out (if they ever were served up much) and times have changed.

Romans introduced most of the herbs we grow in our gardens today, so we don't have to import them. We like, though, to import many spices (like peppercorns, nutmeg, cinnamon and all spice, for instance) that need a tropical climate, but they're not bulky and not much of a drain on fuel to import.

All the same, we should ask ourselves more 'where was this produced?' 'Has this caused problems for farming families elsewhere in the world?' 'Is this food being produced in a destructive way?, 'If so, can we buy better - say Fairtrade or organic?'

Get Into Seasonal Food

Remember I started this in January on a cold, chilly, windswept day. Where I stood is, and has long been, a place for the hearty meaty stew or hot pot, sweetened by winter vegetables like turnips, parsnips, and carrots. Maybe you could throw in some celeriac? Or, you could roast a joint on Sunday with a tasty sauce, or make a meat pie. See my Recipe Section. Save the fat from the meat, and the bone (if there was one) – it's a valuable resource. Make a simple vegetable soup with the remaining uncooked vegetables for a midday meal.

Look up seasonal food for where you live. Old cookbooks are a good source for this, as cooks of old relied much more on seasonal food. My grandfather had an old, tatty cookbook that was my great grandmother's, and it's now mine. It was published in 1911, and it's full of recipes; some curiously old-fashioned and some surprisingly modern. It's called *Coombs*

Unrivalled Cookery for the Middle Classes[3], by Miss H H Tuxford, and it may have come with Coombs baking products. You wouldn't find a book title like that anymore. Maybe you have a cookbook like this in your family?

Get Into Grass-Fed Produce

Farmers and smallholders in the hills have limited capacity to produce plant-based food. Yet, in our wet, maritime climate, we grow herb-rich grass better than most places in the world. Farmers and smallholders can give animals a natural life outdoors, on hillsides where they turn vegetation into nutrient-dense food with next to no outside inputs. This is solar-powered farming, and it's the *Power of Pasture*. It's wildlife-friendly too.

From ancient times of our flint-knapping hunter gatherers (fashioning arrowheads and hide scrapers from stone) to pastoral farmers, this is how we've fed ourselves for thousands of years.

If you think it's difficult to find genuine grass-fed, high welfare produce, you can start by buying local and direct from butchers, farm shops and farmers' markets. In time, supermarkets may respond more to the demand for grass-fed produce.

Wherever you live, there are likely to be food certification labels that show high welfare and grass-fed food that will help. It would pay to dig beneath the surface though, as some labels have proven to be not all that they appear to be. If in doubt, buy local and direct.

---

[3] *Coombs' Unrivalled Cookery for the Middle Classes* (3rd Edition). *Special Chapter on Vegetarian Cookery* edited by Miss H. H. Tuxford, Abel Heywood and Son, Manchester

Quick Takeaways

- Eat more local - shrink your food footprint
- Find out the seasonal foods for where you live
- Use imported food, drink and spices sparingly
- Cook with grass-fed meat
- Get to know your local producers and food certification labels

* * *

Trends in what to eat come and go. But whatever the trends are, one thing is for certain: hills and mountains will always be there. Walk them in the British Isles, and you'll be buffeted by wet prevailing winds coming off the Atlantic and Irish Sea (at least, as long as we keep the climate that way). Tramp along the footpaths and you'll be walking with the animals, just like people have for aeons. Walk south-eastwards with me over more hillsides, onto different rock and soil.

2

Red Rock Farm

Follow me round the compass from where we stand. For our tour we swing clockwise, south and east to a small family farm where my husband, Andy, grew up, north of the town of Kidderminster, just over the Shropshire border into Worcestershire.

I've asked him why they kept only very few cows. He explained that it's sheep farming country, where the short, tufty grass isn't lush enough for farming many cattle but fine for sheep. They hardly ever grew any crops outside of the vegetable garden. Neither did their neighbours, for the terrain is hilly and crops are hard to grow on the shallow, sandy soils.

The pasture knits the soil together, acting as a sponge, holding in water and organic matter, reducing flooding in winter and drought in summer. The sheep poop on those hills, replenishing nutrients that rejuvenate the red sandy soils that overlie buff red rocks. Rocks which tower over the roadside in places, washed and sculpted by the rain into odd shapes. The red sandstone has shaped how people lived and how they farmed.

It's a distinctive landscape. One that's shaped how people lived in more ways than one, as people lived in those rocks until fairly recently. My husband, Andy, used to cycle up to one site at Kinver Edge near the Worcestershire/Staffordshire border, and look down on the farmscape all around – back down to his own family's farm too. They were empty caves at the time of his teenage rambles.

The Kinver rock houses are well known, for it's thought that J R R Tolkein may have taken his inspiration from these houses for his novel *The Hobbit*. He lived in nearby Birmingham while people still lived in them. They were only closed down in the 1960s, but now you can visit them as the National Trust has reconstructed them.

Easter Visit to Red Rock Farm

My husband's small family farm was just to the south, on the same hilly terrain and red sandy soils. It's an environment more suited to sheep farming than growing broccoli. We shall call it Red Rock Farm.

On a warm Easter bank holiday Monday, we were back in the hills. We took a trip up to the Kinver rock houses. We crunched along paths up the side of the 'edge'. It was like walking along a red sandy beach, where trampling feet had worn away the gravel and grass. Sitting with coffee and cake at the outdoor café, we looked down on to lowland to the north. It was a landscape of green and yellow, because most of the fields were under pasture, with a few fields under oilseed rape that we see nearly everywhere in the countryside today.

Later, instead of heading down into the fields of green and yellow, we turned south. We drove past former Red Rock Farm and pulled in off the road. Andy was intent on pointing out the carbuncle of an extension on the side of the Victorian farmhouse.

17

Bolted on since they'd moved, I might add. Their house never looked like that.

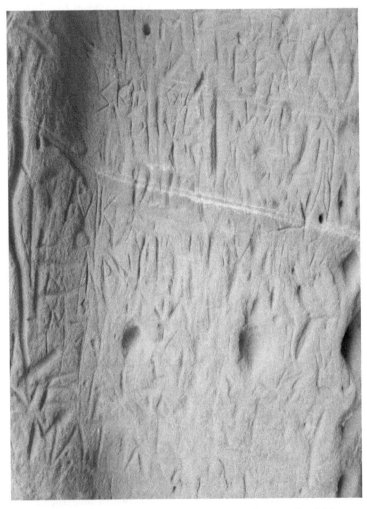

Graffitied red sandstone rocks at Kinver Edge, Staffordshire

Kinver rock houses

My husband, Andy, inside a rock house living room

My mind was more on the lie of the land. In front of the house, the fields sloped gently down to the road. 'So how come you never grew crops?', I asked. He pointed at the slopes rising up behind the house where most of their fields lay. 'Can you imagine getting a tractor up there?' he said. I had no idea. I've never driven a tractor.

It's a factor we might not think about - the terrain and the food-producing environment. In fact, most of us probably barely notice the lie of the land when we drive through the countryside, or look out of a train window. It's not the first thing on our mind, for most people.

That day my husband was in a reminiscing mood, pointing out familiar country lanes. He'd say 'I used to cycle up there to get to the rock houses'. Had we cycled that day, I'm sure we would have clunked down into bottom gear to get up the hills. That's on good touring bikes with low gears too, that can get us up steep hills.

These are not high hills. If you look on a map, the high point around here would be nothing compared to hillier, and even mountainous, parts of the British Isles. But a slope is a slope. Plough a sandy soil and you'll find much of it washed down to the bottom in a rainstorm.

Traditionally, farmers ploughed their fields to sow crops, but around here, even if they converted to no-till or minimum-till (sometimes called conservation till), they'd still need to get farm machinery up on to the slopes. Most of us are not the ones negotiating hillsides (with draft animals) and today, machinery.

Here my husband feels at home, and oddly enough, one of his relatives lived in a rock house nearby. The family apparently has an old tatty photo of him standing in front of his house. He looks smart. He wears a suit with a pocket watch.

* * *

We know little about the life that he, and his family, lived. Today Albert looks out from his house door on a warm, sunny Easter day. He and his wife, Gertrude, do well for themselves, but it's that time of year again when the larder is looking a little thin. He's picked all but a few leeks, and apart from a lone cabbage and some kale in the vegetable garden, there's not much left. Seedlings growing under cold frames have been slow to germinate as the end of February and March has been cold and wet. There's one small cheese left, and some potted meat in the pantry at the back of the rock house where they'd stayed cool all winter.

This is the 'hungry gap' when few winter vegetables are still growing, new vegetables or fruit are barely under way, and new animals are only just being born on the farms below. They, like most people, have been living on home-preserved foods (and some bought), bread from stored flour, porridge and grain-thickened stews.

Gertrude milks two goats, which pick their way around the steep rocky slopes around the house. She makes some cheese from the goats for short-term storage (which makes a soft cheese), but buys in harder cheese for over-winter storage.

* * *

I wonder, what did he, the rock house dweller, think about that pastoral farmscape around him? He would have seen vegetables in his neighbours' gardens, but no large farm growing them. There were no mega-livestock farms either, concentrating animals together in one place. No animals permanently in pens eating soya-based feed.

Instead, our rock house dweller looks out on a landscape more like that re-created at the Kinver rock houses site today. There, the National Trust has reinstated a small area of allotments just below the houses, giving an impression of how this landscape would have looked around here in times past.

Livestock and Feed

Though I saw no farmer further north in Church Stretton to quiz about livestock feed, here I have a farmer to ask. My father-in-law: former farmer of Red Rock Farm.

He tells me that their sheep and pigs were outdoors most of the time, yet, even when they kept them indoors, feed for the sheep included hay from their own hay meadow. They also bought in feed made up of sugar beet waste from a beet factory just to the south (now closed down). The sheep loved turnips too, grown around here on flatter ground, as a cover crop to rest the soil.

You might well say that we should eat the turnips ourselves rather than feeding them to animals. Up to a point, that's true. If, though, I could raise my husband's relative up from the dead and suggest it to him, he might think that's a mad idea.

After all, he lived in a time when protein and fat (barely provided for by turnips) came from the land beneath his feet. Not in a container ship across oceans in the form of protein-rich or fat-rich crops that we can't grow, or find it difficult to grow in this country.

We now grow crops like soya and quinoa in Cambridgeshire and East Anglia, but there's only so much land suitable for growing these new genetic varieties. It's hard to imagine being self-sufficient in these plant-protein or plant-oil rich crops.

Would we want this anyway? Many people value the traditional historic landscape, which, in the British Isles, means cornfields, pasture, hay meadows, commons, ponds for watering livestock, and woodland. The same is true in many other countries.

Returning to livestock feed, the feeding regime at Red Rock Farm was a far cry from the one we read about in the media - a regime where farmers feed livestock on crops which we could feed to people. It's not to say that it isn't a problem elsewhere, but clearly, there are other ways to feed livestock.

So it is that hilly and mountainous landscapes (in cool, temperate climates) lend themselves more to producing animal-based food than plant-based food. From hillside pastures we produce nutrient-dense food: meat, fat, milk, cheese and eggs. Manure replenishes the soils, and we're shod and clothed from the hides.

Nutrient-dense food includes not just meat, but also animal fat. Until recent decades, we've relied on animal fat instead of palm oil, rapeseed (canola), sunflower and soya oils, as we do today. From pasture came fat. Now we cultivate extra land (at home and abroad) to provide the fat in our diet, and fat used in soaps and cosmetics.

We made shoes, waterproof clothes and bags from leather instead of plastic, synthetic fabrics and cotton. Wool off the backs of sheep made our clothes, and their shearing made sheep more comfortable come spring and summer.

This may all seem barbaric, when you think of all the other household bits and pieces that we also made out of bone—buttons and knife handles, for instance. The outer sheaths of horn made cups, and what was left of the bone was ground and added to clay to make bone china (porcelain). Now we use other resources (often plastic), but all those resources have an effect on our environment.

A Hardy Breed

Hill farmers are a hardy breed. They're dealing with challenging terrain, harsher winters and a short growing season. Hence, they tend to favour hardy breeds of animal.

At Red Rock Farm, they mainly kept Swaledale sheep - a breed which belongs to the Blackfaced Mountain family. They have black faces with distinctive white markings, and a Roman nose. They're hardy: at home in the cold, wet, bleak conditions of the Yorkshire Dales, but we've moved them across-country to places like Red Rock Farm, where they seem to prefer to be outdoors for much of the year, come rain or shine.

Sheep breeds, like Swaledales, are part of our farming heritage. They're native, rare breeds, and it is often hill farmers who are the rescuers, bringing endangered and rare breeds back from the brink. Genetic diversity has developed over generations of livestock living alongside people. We may not know it, but our lives have been closely entwined.

Small-scale farmers are often closer to their animals, knowing them individually, and knowing their ways. This way of life on a small family farm is a long way from the confined feedlots where animals are confined in pens, eating formulated rations, never getting the chance to tear and munch at grass and wild plants, with birds circling overhead.

Yet, these farmers could become an endangered breed themselves. They're facing the effects of owners selling off their property for holiday cottages, and their low incomes. Andrea Meanwell in *A Native Breed: Starting a Lake District Hill Farm*[4], wrote in 2016, that the average income of a hill farmer is £8000 per year. At the time of writing, that's just under 10,000 US dollars.

Excerpt from *Hefted*, by Andrea Meanwell:

> *We teach our children the way to walk,*
> *To feed, observe, and care for the sheep.*
> *Feet planted where ancestors trod*

---

[4] *A Native Breed: Starting a Lake District hill farm* by Andrea Meanwell, Hayloft Publishing

As they became hefted to the fell for life.

Hefted, anchored, rooted, tied and bound,
Sheep and shepherdess together,
Generation after generation walk the fell,
We're ring-fenced, hobbled, grounded, home

With kind permission of the author and publisher of *A Native Breed: Starting a Lake District Hill Farm* (Hayloft, Cumbria: 2016)

Alongside those hardy breed animals, and hardy breed of farmers, lives wildlife that has almost disappeared from modern farmland. My mother-in-law tells me that on Red Rock Farm, alongside sheep and pigs lived these birds: red-legged partridge, grey partridge, nuthatches, goldcrests, spotted and green woodpeckers, grey wagtail, pied wagtail, siskins, jays, woodcocks, song thrushes and mistle thrushes, barn owls and little owls. Some of these birds are very rare now. They had orchids in the meadow, and they had a survey done once that recorded 80 species of wildflowers. That's wildlife-friendly farmland.

What Can We Do?

If you live in, or close, to an upland environment, support hill farmers by buying their produce and championing their cause. They're the keepers and carers of our rare native breeds. Custodians of our countryside.

Cut out the Middle-Man or Woman

How on earth would you know whether you're buying meat, butter, cheese, milk, eggs, or other products that have been farmed in a sustainable way? You can be more sure by buying from a farm shop, farmers' market or independent butchers. Prices aren't necessarily high, as you will pay farmgate prices. If these places are hard for you to get to, you could pay for delivery. Here in the UK, more farms are selling direct to the consumer again, and even more would do so if we supported them.

The advantage of buying from farm shops, independent butchers or grocers, is that if they don't sell it, but you ask for it, they're more likely to listen. You're one of a small customer base, and your words and requests mean more to them than to a supermarket buyer.

Think More Hogget or Mutton

Food, animals and crops were a precious resource in the past, and we wasted nothing. Up in the hills, most sheep would have lived longer lives, and we would have been eating mutton from mature animals, or hogget from animals that have had two springs out on grass, so they're not quite at maturity.

We're more used to eating lamb these days, as the meat is more tender and easier to cook. Remember that hungry gap. If you live in the global north, traditionally, lamb would not have been seasonal food at Easter, at the time we visited Red Rock Farm. They would only just have been born. Now we see lambs born in winter to sheep who have been artificially inseminated and fed on concentrates, not grass. Not only is this not very sustainable, it's harder on the animals, and they're not

27

as healthy on this kind of diet. Neither is the meat as healthy, because grain-feeding affects the balance of fats in the meat.

Eat mutton or hogget instead, if you can get hold of it. This might be a challenge, as the sale of mutton and hogget has been in decline for decades, or at least since the late 1960s, judging from a cookbook I have that belonged to my parents where mutton features. See my Recipe section. You need to cook longer and slower than you would for lamb or cuts of meat that we are used to today.

Look out for Rare-Breed Meat and Dairy

You may find farmers near you are selling meat from rare breed animals, as they are increasing in popularity. More farmers find that keeping animals that are tough, hardy, and more content outdoors keeps them healthy and avoids having to buy in commercial feed.

They happily graze on rough pasture, which is more wildlife-diverse. This meat has a 'terroir' (taste) that reflects the pasture on which the animals feed.

Dorothy Hartley, who wrote *Food in England: A Complete Guide to the Food that Makes Us Who We Are*[5] describes how the food that an animal ate flavoured a joint of meat. Marsh mutton, which comes from sheep that have grazed on salt marshes, has a slightly salty flavour, and cooks flavoured it with hot larver sauce, made with the larver weed that grows on the marshes. We served Midland mutton, from sheep that have grazed in orchards, with fruit sauces.

---

[5] *Food In England: A complete guide to the food that makes us who we are* by Dorothy Hartley, Piatkus (2009 edition)

It may seem wrong to us today that we should relish that the meat of an animal has a terroir or specific taste. Dorothy Hartley, writing in 1954, said that we'd lost a sense of the connection of meat with an animal, and we might say that many today are connecting meat with animals, and are repelled. Yet, are we repelled at the thought of what wild animals would taste like: those trying to eek out an existence in a modern, intensively farmed landscape? Would they taste of chemicals: the fertilizers that have replaced animal manure; the pesticides that substitute for biodiversity in pest control?

Eat More Nose to Tail

In farm shops, at farmers' markets and at independent butchers, you will find more slow-cook, thrifty cuts of meat too. Think middle neck and scrag end of lamb (or mutton) or pork shoulder. We probably think scrag end of lamb or mutton is poor fare, but slowly simmered in the pot, with the last of the winter vegetables, and a few emerging spring vegetables, this is good food.

Also featuring in my dated and dog-eared parents' cookbook are recipes using offal, or organ meats. Steak and kidney pie, liver and onions. What might be a stretch for us now is tripe pie, mock hare (calf's liver), pig's trotters, and stewed calves' feet, featured in my great grandmother's cookbook. Whether these foods will become staples in your kitchen really depends on how nose to tail you're prepared to go.

My sister-in-law, who grew up on Red Rock Farm, loved tripe, onions and eggs, as her grandmother used to cook it for her. The rest of the family could never understand it. She was on her own there (with her grandmother).

29

Look for Naturally Processed Food

Look out for small-scale producers of cheese, yoghurt, sausages, and hams for instance.

They're more likely to be using local produce, processed in a way that uses fewer resources. This is healthy food, that in the past would have got us through that hungry gap. Think of those mature cheeses kept cool in cellars, along with naturally-aged sausage meat. There's a reason why these foods are found around the world in traditional food cultures.

Quick Takeaways

- Eat mutton or hogget, instead of lamb, if you can get it
- Slow cook more thrifty cuts of meat
- Get back into organ meats (offal)
- Add in a little respect for the Hungry Gap: seek out cheese, yoghurt, sausages and hams that are naturally preserved
- Raid your family archives for cookbooks and recipes

* * *

You may not live remotely near any hills. In which case, we need to come down onto the lowlands and explore further. Exploring why ancient ways of eating could help combat climate change is where we're going next.

3

Cow and Corn

We start from where I left off, in hilly country on the Worcester shire/Staffordshire border. We now head round the north of Birmingham, onto flatter ground where corn is always rotated with livestock.

These are farming ways that have been in existence for thousands of years, creating our historic landscape. It's a theme I keep coming back to in any writing about food.

This brings me to archaeology and ancient food. You might wonder how that could have any relevance to food that could rightfully sit on the countertops of your kitchen today? Look at what people ate in the past, and you might see what I mean.

Archaeology, Ancient Food and Climate Change

Dig down to find the remains of past farming lives under the plough soil. Year after year quarries expand, housing estates appear, and flood alleviation schemes criss-cross the countryside, exposing food remains surviving in hearths, corn dryers, pits and ditches. They sit alongside dark marks left in the soil by houses and field systems: the remains of houses, ditches, yard surfaces and field boundaries.

We've come down from the hills, and now we're on lower ground. Looking northwards from Birmingham. Beneath the newly-built supermarkets and retail parks in Stafford and Lichfield are the remains of burnt-down buildings, ovens and hearths that take us back to Saxon times and the Middle Ages.

We take a turn into Stafford and go back in time to 960 AD. On St Mary's Grove near the centre of town there's a woman working by a kiln or corn dryer. She's carefully scooping out of the last of the corn dried before it goes into storage. She then rakes out burnt dregs from the flue. Blackened grains of rye, oats, a few barley grains, chaff and straw still line the floor of the flue. Smoke rises from other ovens nearby.

Local Crops for Local Soils

Not so far away, farmers are bringing crops in from surrounding fields, haggling with buyers on the market. They would get the best price for wheat, but what they have more of is oats, rye and barley. They know these crops grow best on the poor soils which many of them farm. These are not poor, hapless peasant farmers though. They know their land and what it can produce.

They rotate cereals with livestock, whose manure, along with nitrogen from peas and beans, beef up the nutrients in the hungry soils. Dung and any rotting organic matter are needed to feed the grain.

And, along with the dung came dung beetles and dung fungi, adding to the biodiversity and nutrients in the soil. No fossil fuel-based chemical fertilisers are known at this time: it's dirt that farmers rely on.

Nevertheless, they still grow crops that take fewer nutrients out of the soil than wheat. Locals also know that in cold, wet winters, oats will thrive better than wheat. They don't need to be told to keep nutrients in the soil, minimising carbon dioxide and methane emissions that today contribute to global warming. They just do - they have to.

A Fire in Lichfield

Around the same time, in Lichfield, to the south-west, a fire breaks out. Within sight of Lichfield Cathedral, a building is ablaze. The next day, the building collapses, and the remains of the thatched roof cave in. Among the smouldering embers are blackened grains of rye, a little wheat, chaff, and tiny seeds of weeds that have grown in surrounding fields. These are all that remained of the thatched roof.

Roofs around here, in Saxon Lichfield, are thatched mainly with rye. Locals grow a mix of rye and wheat, but rye is always dominant. It's a mix that works well on surrounding nutrient-poor light, sandy soils, rotated with livestock and beans to keep up soil fertility.

They know that diversity pays, for there's a better chance that one crop will succeed, whatever the prevailing conditions that season.

Food miles are low, and they've used no fossil fuels in bringing crops into town.

Oats and Oatcakes in the Potteries

Let's head up to Stoke-on-Trent and fast-forward in time to the 18th century, winding our way past bottle kilns of the newly burgeoning Potteries. We can sample Staffordshire oatcakes that are the fast food of factory workers. Oats are a crop that win over wheat in cold temperatures, on hilly ground and poor soil of the Staffordshire Moorlands and foothills of the Peak District to the east.

Heavy clay soils here lend themselves more to pottery making and pastureland than tilling for crops. The pottery industry mushroomed from a small cottage industry of artisan potters making butter pots for dairying, which reigns on damp clay soils.

My mother's family came from here and around Stafford. My brother and I, with the neighbours' children, used to play in a farmer's field. We weren't trespassing because the farmer was happy for us to play there.

It was a steep-sided paddock used to graze horses, sloping down to a trickling brook. We used to wade around in wellies, getting ochre-coloured hands because of iron that seeped out of the ground, covering the pebbles in the stream. We used to dig out stiff clay from the stream bank to make clay figures and try unsuccessfully to bake them in the hot summer sun. Our parents and grandparents told us that summer sun wasn't enough - we needed a kiln. But, being children, we just had to try. It took effort to dig out that clay.

When our artwork eventually cracked and fell apart, there were a few fleeting, trembling lips, then all was forgotten as those summer days were just the best.

It took effort too for farmers to plough and plant crops on

those stiff clays, so it's no wonder that this area has always been mainly dairy country, under pasture. Oats, for oatcakes, were served to factory workers from house windows as they walked along the street. They were usually served with savoury foods, and went well with pastured cheeses.

Potato Famine Refugees Come to Stafford

Back in Stafford, in the year of 1851, Bernard Cavanagh has just answered the door to a census official. He gives his name, and that of his wife Susan and family. His birthplace–Ireland. He's been here for a couple of years or more, living on the outskirts of town, arriving during the later years of the Irish potato famine. Locals are getting used to the Irish, and in that community, Bernard and family are settling in. Bernard is now making a living as a groom.

Did he come to England as a starving potato crop farmer? Or was he just another fleeing from the economic downturn that went along with the disastrous potato blight in Ireland? We don't know. But the Irish situation was different to the farming system that Saxon farmers around here once knew.

Bernard fled a disaster that need not have happened, had the humble potato not become a staple of the poor. A monoculture of the Irish Lumper potato. The one crop that would succeed on poor, wet land to which impoverished peasants had access. Grain and beef were exported to Britain by English absentee landlords, who owned much of the better farming land, so food wasn't in short supply.

Food security failed in Ireland during these years.

Archaeologists Come to Stafford

It's the early 1980s: archaeologists are excavating in St Mary's Grove, where a woman once raked out blackened remains from a kiln. They've uncovered the kilns and taken soil samples that go to the University of Birmingham, where an archaeobotanist, Lisa, records the charred and blackened crop remains.

In the late 1990s, another archaeologist from Worcester follows on from this work. She's called in to take samples from a wet, boggy hole sunk into the peaty ground of King's Pool, just to the west of the ovens on St Mary's Grove.

She meets up with James, who'll study the pollen from these samples, and they both climb down into the deep, shuttered hole in the ground with sample tubs in tow.

The signature from the waterlogged remains from that hole tells a very similar story to that from the ovens in St Mary's Grove. Only, the diverse crops brought into the town are there in the form of pollen and seeds which had blown across the marshy ground and settled into the sediments. Now, the backfilled hole lies somewhere under the car park of an Asda supermarket.

And Today....

The archaeologist from Worcester who arrived in the late 1990s was me. Although I didn't know it at the time, Bernard Cavanagh and family living on the outskirts of Stafford in 1851 are my Irish ancestors. They may have been economic migrants, coming into Stafford during the later years of the Irish potato famine, giving me a link with the town.

In the years since, with my microscope, I've scanned through charred grains of oats, rye, barley and wheat (but no potatoes) from this area. The pattern of their distribution in the farm-scape, in relation to soils and terrain, has imprinted itself into my brain. It's a familiar pattern.

I see a general scarcity of charred cereal crop waste. Elsewhere to the south, that waste is abundant in the landscape (you can travel there in the next chapter). But here, through the ages, we find it only where there are concentrations of people. It's an area where people have always cultivated and processed crops conservatively, and raised livestock. That's what worked, and it should still work today.

The important topics of climate change, environmental degradation and food security revolve around in the media. It's had me thinking there's much we can learn from the past. In fact, the past has lessons for us today that we can use for the future.

Lessons From the Past

What could we learn from Saxon peasants and Irish immigrants? That we grew local crops for local soils. We farmed in a way that suited the landscape, and in most years diversity worked and fed people. Unless, of course, an elite land ownership prevented it from doing so.

The solid rock under the feet of peasant farmers was not going anywhere. Neither is it today. It will always weather out into the same soils. However much we believe in the silver bullet-the one diet that will save the planet, there really isn't one. So much depends on the soils.

Back in the day, we would not have seen wheat, wheat and more wheat. Rather, oats, rye, barley, cows, pigs, flax, hemp, horse beans, peas and vetch. Today we override limitations of soils and climate in so many ways, but we're paying the price.

We grow relatively nutrient-hungry wheat with chemical fertilisers, produced from non-renewable fossil fuels. We release **methane** and **carbon dioxide** from their production, and from depleted soils. Those greenhouse gasses are implicated in global warming, but where food and farming is concerned, it's mostly blamed on livestock.

We need to go back to feeding the soil with animal manure, and growing more suitable crops. For decades, we've imported huge quantities of wheat into the UK from Canada, Russia and the US, transported by burning fossil fuels. None of this makes much sense in a world of global warming and peak oil.

Manure, not chemicals, is what we need, from animals who have had as natural a life as possible out in the fields for much of the year. Well-raised, free-range pigs will have had a good life out on grass, sometimes with the opportunity to root around in woodland. Pigs were outdoors for much of the year on my in-laws' farm, with small huts for shelter, like they still are on most small farms where there isn't the infrastructure to keep them permanently confined.

They will have manured soils in much need of organic matter, not chemicals. If it were not for UK legislation, we could also feed pigs on food scraps and food processing waste just like we used to. Apparently, they love whey, a by-product of butter and cheese making.

Climate Change and Food Security

Ancient food and climate change are getting a glimmer of interest on social media. If you're a foodie and want to help with your food choices in the battle against climate change, it should give you hope. BUT, mostly this subject revolves around the idea that climate change (seen mostly as global warming) is happening. So, we need to bring ancient crops from warmer climates into colder regions.

The thought seems to be that, as global warming is happening, let's prepare to grow older Mediterranean crops in our cooler climate because that's where we'll be in a few years time – living in a Mediterranean climate in the north. We'll need the genetic diversity in those older crops, and their drought tolerance. So, get with it and adapt!

I have a different take on this, though. My take is that if we support farmers in growing food according to local ancient farming traditions, we can not only help to prevent climate change, but also environmental degradation and loss of food security. All by what food you bring into your kitchen and cook with.

Those ancient farming traditions we saw around Stafford and Lichfield conserved soil, kept carbon (and greenhouse gasses) in the ground more, and gave people a degree of food security. The same has been happening the world over, and as much of the world is a patchwork of fertile soils and less than fertile soils, these farming ways have stood us in good stead.

What Can We Do?

For those living in a cool climate, why not:

Bring in Some Rustic Traditional Foods

On a cold winter's evening, some warm, filling and inexpensive food wouldn't go amiss. Most of us like porridge and pancakes today, only we might want to take a tip from our grandparents', or great grandparents' book, and soak the grain or flour, or even ferment it. See my recipe for Staffordshire oatcakes (like pancakes) in the Recipes section.

The advice to soak porridge oats at least the night before cooking was on our porridge oat packaging until very recent times, but now we rarely see it. Old processes have common sense behind them. Soaking and fermenting reduces compounds (like phytic acid) in the grain, which combines with vital nutrients in our guts and prevents them from being absorbed. Unsoaked or fermented grains, essentially, rob us of vital nutrients. We used to soak grains in water, buttermilk or whey.

Do the same with pearl barley and throw it in your casserole or soup. Try a little rye bread.

Above all, attune more to a medley of food that is closer to an ancient food culture. If you live in temperate climes, where there are cold winters and poor soils, this is the makings of a place-based diet. These are only simple takeaways, but once you embrace these, I'm sure you'll want to dive in deeper.

Delve Into Your Family Archives

When I went to my grandparents' house, on a rainy day, I would either be rummaging in boxes looking for Nan's unravelled wool to knit with, or sitting with my grandad looking through old photographs and keepsakes.

They were in a small suitcase, and he would go through them with me, telling me who was who. He would pull out carefully folded business stationery and invoices that belonged to his father (my great grandfather), who ran a small house-decorating and signwriting business.

What about your family? Perhaps you had, instead, a farmer,

baker, dairy maid, brewer, or grocer in the family? You may not know, but it can be surprising who can emerge from your aunt's dusty attic once you ask, along with their everyday effects. Their stories might be a springboard for connecting with older and different food cultures: perhaps cultures which challenge our worldview in a beneficial way.

My great grandmother's cookbook (*Coombs Unrivalled Cookery for the Middle Classes!*) came out of the archive case too. You might find too that you have potential Irish potato family refugees lurking in your midst, like I did.

If your family were the opposite of hoarders and you're left with little to go on, try online family search sites. Census records, birth, marriage and death certificates can hold so much information.

Quick Takeaways

- Eat porridge oats and oatcakes
- Thicken soups and casseroles with pearl barley
- Bake some rye or spelt bread
- If you're a west midlander, how about some pork from free-range Tamworth pigs (a breed from Tamworth, near Lichfield)?
- If you live in a warmer climate, find your equivalents. How about Durum and spelt wheats, rice, millet and sorghum, and local ancient breeds of animal depending on where you live?

* * *

You might think ancient grains now have a middle-class cachet to them, but perhaps they ought really to be the food of the masses in areas where they have been grown for thousands of years? Ancient food brings me on to another subject: artisan food (another class of food which now has a middle-class, foodie air about it).

4

The Breadbasket

We've toured just to the north of where I live, where we're rooted in a largely pastoral culture that's lasted for thousands of years. Nevertheless, there's a landscape which lends itself better to a more plant-based diet, though never totally. Even though soils are more fertile, and growing crops in bulk is easier, livestock have still always been a part of the landscape.

Let's take a turn south now on a tour around that diverse landscape. We go through south Warwickshire and south-east Worcestershire. Think William Shakespeare (Stratford-upon -Avon) and Inkberrow (home of the BBC Radio 4 series, *The Archers*), and head south.

Ridge and Furrow, Barns and Windmills

Here, there are many more large wheat fields and, today, fields of rapeseed, colouring the landscape gold and yellow in summer. It's a landscape of cornfields: much more than we find further north. There are old windmills, watermills and tithe barns. They're the legacy of a well-organised, corn-growing culture.

That culture, in this area, has roots that go far back into prehistory.

This is what we could call the breadbasket of the West Midlands. It's only a fairly small part of the region, and it's also true of much of the country. I've seen the claim that only around 30% of the British Isles is made up of fertile land that is ideal for growing crops, and my estimate is that it's similar for the West Midlands.

I don't know the source of this claim, but it doesn't surprise me, considering what I know of the region where I've lived and worked for years. The main wheat-growing regions in the British Isles lie east of a line from Hampshire to Durham. West of this line, it's damp, rainy countryside, and wheat doesn't always fare so well here.

Warwickshire Windmills

I cycled around the eastern part of this area, some time ago, with a cycling friend on what we called our 'Warwickshire Windmills' tour. Everywhere, I saw a type of landscape around me that I recognised from old maps, but there was more of it here than I'd seen before. I shouted out to my friend, 'Hey, look at all this ridge and furrow!' It all ties in with the windmills. It's really old, you know!' But, I think my words were lost in the wind, as Dave replied 'Hey what? I think we turn left here'.

There are hints of an older landscape, now grassed over. If you go off the beaten track here you will see many fields that look like this:

Littleton Pastures at dawn. Copyright Aisling Nash, Ashtree Heritage[6]

No wonder, as lumps and bumps in fields were what we saw. Ridges made by Saxon and medieval ploughs turning the sod sideways, forming a ridge and then a furrow. If you look carefully you can see a slight 'S' shape at the edges of the field where the oxen swayed round to turn the plough.

Look again, and you'll often find more lumps and bumps in the grass nearby. Sometimes you'll see a local history and archaeology group out surveying these lumps and bumps which are deserted medieval villages.

It has been a curious mixture of culture and fertile soils that has created this landscape of cornfields, barns and windmills.

---

[6] Ashtree Heritage: https://uk.linkedin.com/in/aisling-nash-18b51535

Me at Napton on the Hill windmill in Warwickshire: on a cycle tour

Littleton Pastures in the mist. Copyright Aisling Nash, Ashtree Heritage

Summer in the Cornfield: Corn Dryer Central

Summer is nearly past, and the corn is gathered in. We're heading towards autumn and it's time to reap the rewards of the harvest. In medieval Warwickshire, it's been a busy time – many hands have been at work in the fields, and they still are. Damp grain will spoil, and there's lots of it, so a dedicated band of farmers get to work on drying a batch ready for storage.

Further north, we came across a woman raking out burnt grain from a corn dryer in Saxon Stafford. There, corn dryers are rare, and mostly we find them where there's concentrated settlement.

Here, a farmer stokes up a corn dryer in a field, fills it with grain and leaves it to gently toast, while he attends to other jobs nearby. Soon, though, there are shouts as flames come billowing out of the dryer. The grain has caught alight, and nothing can be saved. He rakes a burnt dump of grain into a nearby ditch, to join others from previous years. It's collateral damage in a landscape used to drying and storing grain in bulk.

* * *

Centuries later, we can find burnt grain under the ground surface anywhere around here. Here, there's that burnt grain dump, still there in a medieval farmstead ditch. Further south, towards the Gloucestershire border, in a quarry at Severn Stoke, archaeologists unearth blackened grain in grain storage pits, undisturbed since prehistoric times. In the next field were small four-poster structures that were granaries.

We also have a better chance of a 360° excavator scraping the stones of a Roman or medieval corn dryer, before an archaeologist can say ''Whoa', here more than anywhere else. These distinct burnt dumps of grain in ditches and pits are a snapshot in time of crop processing gone wrong, or maybe just 'spring cleaning' of a corn dryer.

Near villages, large tithe barns that held those grain harvests are still there. You can visit one at Leigh, just south of Worcester. Windmills stand out on higher ground on the horizon here in Warwickshire.

Hard-Pressed Land

Growing and processing grain in bulk still carries on today. Breadbasket country anywhere in the developed world is pumped hard to produce grain. Only now, we mine the soil by using chemical fertilisers. We push out wildlife (called crop pests) and eradicate disease with more chemicals.

Farms are becoming more specialised. Where once we had mixed farms, we now have either pastoral or arable farms. Even so, in this area, where soils are relatively fertile, people always kept livestock. This much must be true, as the bones of farm animals lie under the ground alongside the dumps of burnt grain.

People still needed nutrient-dense food, full of protein and fat. They also still needed to keep soils fertile. Without livestock to manure the land, soil fertility would eventually diminish. That's even while growing nitrogen-fixing beans and peas in the mix.

Only now, as we hear warnings that we have between 50 and 100 harvests left in the soil, are some re-thinking post-war farming.

We must adopt a more plant-based diet and embrace 'unprecedented change'. We see and hear the advice in the media often at the moment. We're conscious of the amount of meat we're eating and how we're producing it.

But, we're less conscious of where the plants for our more plant-based diets will come from. And, the environmental footprint of this food.

Margaret Atwood said 'All bread is made of wood' in her poem

'*All Bread*'[7]. It took trees from the landscape, chopped by an axe in early times, to make that bread. It took the death of small animals. Those that couldn't live in the cornfields because cornfields are not their habitat. And, those that could (that died naturally) and became part of the cycle of life: decaying into the soil, feeding the corn.

We could never take sufficient plant-based food for granted in the past, and nor should we now.

British Flour: British Bread?

Breadbasket enclaves everywhere grow so much wheat, and I've often wondered where it all goes. I hear we import much of our wheat for bread-making into Britain. So, what happens to our locally-grown wheat?

We hear most about how we grow wheat for animal feed, which is an example of post-war over-production and our need to use up the grain mountains of the European Union and the rest of the developed world. But, what about wheat grown to make bread?

Back in 2009, the daily newspapers announced that the first British loaf of bread was on sale, made by Hovis. The first one since the repeal of the Corn Laws in 1849, after which tariffs and trade restrictions on importing grain were relaxed. It seems a wild claim, but there might be something in it. At least, as far as commercial supermarket bread goes.

I've just ordered some flour made from rivet wheat; a wheat

---

[7]   "*All Bread*" from SELECTED POEMS II, 1976-1986 Copyright © 1987 by Margaret Atwood. Reprinted by permission of Houghton Mifflin Harcourt Publishing

once common around the English Midlands. Like most old English wheat, adapted to our mild, damp climate, it doesn't make a lofty loaf, but would make good biscuits. It isn't suitable for the commercially-made bread that you see in supermarkets.

For this, we import huge quantities from Canada and Kazakhstan. Even so, the wheat variety Hovis is using is a new strain that suits the climate, but also a modern bread-making process called the Chorleywood process. This crunches the rise time (proving time) down and uses a slew of flour improvers and chemical additives.

The traditional long proving time is when the process releases nutrients and uses up gluten, starch and anti-nutrients. So, what we end up with is modern bread that gives us fewer available nutrients, and more gluten, starch and other chemicals to digest. No wonder that researchers are questioning whether modern bread is good for us.

Hopefully, the situation has improved in recent years with the surge in interest in sourdough and other artisan bread types. We still have working mills in the UK selling locally-grown and milled flour. The Sourdough School[8] publishes a list.

In the UK, Bakery Bits[9] sells flour from heirloom grains, grown by various British producers, and in the US, Community Grains[10] sells *Identity Preserved* products which are fully traceable from seed to table. There are 23 points of identity, including the farmer, the variety, land quality, biodiversity, water use, and the type of mill used. And so on.

---

[8] The Sourdough School; https://www.sourdough.co.uk

[9] BakeryBits – Artisan bread baking equipment; https://www.bakerybits.co.uk

[10] Community Grains; www.communitygrains.com

Christine McFadden, in *Flour: A Comprehensive Guide*[11] shows us that people from all walks of life are discovering terroir in wheat, just as wine has terroir. Read more about terroir in the next chapter.

Ancient and Heirloom Grains

As the heat of summer wanes, at harvest time today, there are small pockets of land where wheat stood tall in the field all summer. It's almost to your shoulder, unlike modern wheat which is barely knee-high. There are a few farms growing ancient and heirloom grains which we've grown in this country for hundreds of years or more.

Ancient grains, like spelt and einkorn wheat, have been little changed by selective breeding. They're more or less just the same as the grains eaten in the past by our prehistoric ancestors.

They were also dominant crops further back in the past. Spelt wheat was the dominant wheat through much of prehistory and into Saxon times in Britain. Einkorn was never dominant here, although it has been in the Middle East and much of Europe.

We could consider heirloom wheats to be ancient grains, but, think of them as more like an heirloom you would hand down in your family. Rather than a familiar piece of furniture or a christening shawl knitted by your great-grandmother, it's a crop. A carefully tended crop, re-sown every year and passed down through generations.

I've been growing an heirloom wheat on my allotment: Welsh April Bearded wheat. I also have some Hen Gymro to sow,

---

[11] *Flour: a comprehensive guide* by Christine McFadden, Absolute Press (1st Edition)

another Welsh heirloom wheat. They're more genetically diverse, lower yielding, but take fewer nutrients out of the soil.

Think of pockets of ripening golden corn appearing in summer, countrywide. Blue Cone, Devon Bearded, Hen Wenith Coch, Old Welsh Hoary, and Old Cumberland, are a random smattering of English and Welsh heirloom wheats, ready to fold into bread, pastry and biscuit dough. You'll stomach them better, and so will our landscape.

Yields aren't everything: the soil can only give so much. More resilience, in the face of changing weather and diseases, season by season is worth more. Thatchers would welcome the change too, seeing as the long straw traditionally used to thatch roofs (there are still plenty of thatched houses around here) can be hard to get hold of in some years.

What Can We Do?

You could:

Bake Your Own Bread

Today, we import much of our wheat for bread making. Most of the bread we buy in the supermarket is pap, filled with preservatives and flour improvers. The rapid fluff-the-bread-up-quick baking process leaves behind undigested gluten and starch.

If that doesn't leave you enamoured with your daily loaf of bread, then maybe baking your own would. Simply baking using the recipe on the back of a flour packet provides you with bread that has either limited, or even no flour improvers and preservatives, and bread without packaging.

Homemade bread made with traditional sourdough or poolish methods are better for you, and are cheaper in price than shop-bought. Poolish baking uses tiny amounts of yeast (which can be fast-action dried yeast) with a long proving time.

Buy From an Artisan Bakery

If you feel you don't have time, support a local artisan bakery. They used to be everywhere. Despite baking with locally-produced flour, and offering slow-risen and sourdough breads, traditional bakeries never used to be called 'artisan'. Now they are and they're a rarity. Sure, it will cost more, but in the next chapter, you will read why cheap food is not really cheap food. You're paying in ways (in real money) that you might not realise.

If you bake bread, pastry or biscuits with grains grown locally, or buy from an artisan baker, you'll be helping conserve soil and cut emissions associated with grain imported in bulk over thousands of miles. Another gain is that you help to keep traditional baking skills alive.

Eat Bread Made From Heirloom Grains

Eat bread, too, made with heirloom grains as they're better adapted to local conditions, and are well-known for not needing molly-coddling with extra chemical fertilisers and pesticides.

The chances are, too, that if you're wheat sensitive (not-coeliac) you'd tolerate ancient and heirloom grains. Gluten may not be the problem. The gluten and starch content and make-up of these grains is different, and many people find they tolerate them better. Modern breeding of cereals has focused on getting high yields and a dough that rises well, but this doesn't necessarily agree with our constitution. Food baked with heirloom grains is also more nutritious. It's all in the make-up of the grain, the milling, and the baking process.

Remember too, whole grains (pearl barley and other soaked grains) in soups and stews mentioned in Chapter 2.

Dig Deeper Into Your Local Landscape History

Traditional ways of farming are often more suited to the local landscape, and so produce good, sustainable food. These ways are etched into the landscape. If you want to go further, they are to a degree, still there for you to find. They are your clues.

Look at old maps of where you live, and you may see field names like 'Oaten Field' or 'Ryelands'. They might conjure up images of food that could be on your menu more.

You might have a local archaeology and history group near you who would welcome a hand.

Secondhand bookshops, or antique shops, are worth a peruse. If you have the time, libraries and local record or archive offices are a treasure trove too.

You might not have to go far at all. If your family is local, who knows what might be in your family archives? Even if you've looked through them before, after reading this, you might look at them again with a fresh eye.

Quick Takeaways

- Learn how to bake a sourdough loaf, or a slow-rise loaf with yeast (poolish bread)
- Try heirloom grains
- Look for food certification labels like 'Conservation Grade'

or Soil Association Organic on flour, found in the UK (look for equivalents if you live elsewhere)
- Dig into your family archive again

Man shall not live on bread alone, Jesus said in the Bible – *but by every word that comes from the mouth of God.* We, in this case, shall not live on bread alone, as in this story we move on to a landscape of market garden vegetables, orchard fruits, cider, perry and hops for beer.

5

Terroir: Vegetables, Cider and Beer

We swing slightly south-west now into a market gardening area that came into being in the early 18th century. It fed towns and cities like Worcester and Birmingham on its doorstep. Here, in the Vale of Evesham, we walk mostly on fertile soils.

I've wondered why this area became such an important market gardening area. I thought it was because the soils are generally fertile, but it seems that not all of the success of this area was down to fertile soils. It was partly that this became an area of smallholdings - small market garden farms rather than large arable farms.

Smallholdings and orchards were everywhere until fairly recently, rising to a peak around the time of the Second World War.

My dad (with my aunt) was evacuated from Wimbledon in London to a village near Evesham at this time. He was very young, and all he remembers about that time was that they stayed with a family that lived on a smallholding. Somewhere, we have a photo of my dad and aunt as small children sitting on an orchard wall.

I thought I could work out where they were by looking on old maps for a parcel of land of the size of a smallholding with an orchard. The trouble is that much of the village was like that. I don't rate my chances of pin-pointing where they were.

For instance, if you were in the village of Badsey, near Evesham, around 1911, you would have found the whole village thriving with smallholder industry. The Badsey Society found that more than 80% of householders registered themselves as market gardeners in the 1911 census. You can find out more in *The Last Market Gardener Project*[12].

They grew flowers as well as vegetables for market amongst the orchard trees, keeping soils fertile with manure from sheep and cattle. It was a landscape that supported diverse wildlife, but after Second World War it saw many changes. Hear mother owl's story[13] to find out what has happened to wildlife and the soil. And, how you can help to support a return to more nature-friendly farming amongst the apple trees.

If you live in this area, a project up and running at the time of writing is the 'Market Gardening Heritage' project[14] which aims to survey and restore small market garden buildings called hovels, and to record local memories which will be included in a reminiscence pack.

---

[12] *The Last Market Gardener Project*; https://www.badseysociety.uk/market-gardening-and-farming/last-market-gardener

[13] *Support Nature-Friendly Farming to Save Mother Owl* on my blog; https://lizpearsonmann.com/2019/08/nature-friendly-farming-and-owls.html

[14] *The Market Gardening project*; https://www.explorethepast.co.uk/project/market-gardening-heritage

Autumn in the Orchards

So, this is where we are now. The number of orchards has dwindled and smallholdings are rare. CPRE The Countryside Charity, has produced a guide called *From field to fork: The value of England's local food webs*[15]. They say that although more than 2,300 varieties of apple originated in England, only one in three apples we eat is grown here. In the last 25 years we have lost half of our orchards.

> *It's a sign of our increasing separation from nature*
> *that we are losing sight of where food comes from*
> *and how it is produced.*
> Monty Don: From CPRE, *From field to fork: The value of*
> *England's local food webs*

It's early autumn: apples, pears, plums and damsons are still being harvested. Animals of all kinds make off with some of the windfall in gardens and in smaller, less commercial orchards. As autumn progresses, apple-pressing events pop up around here run by Transition Town[16] groups – the reclaimers of local economy. We could be eating some of the stored apples as toffee apples on bonfire night – make it a local one!

---

[15] *From field to fork: The value of England's local food webs*; https://www.cpre.or g.uk/resources/from-field-to-fork-2

[16] Transition Network; https://transitionnetwork.org

Stewed and Slowly-Baked Sweet Delights

Let's pop into Pershore, the home of the Pershore plums. As autumn gets cooler, here, you can eat a Pershore plum pudding and drink some plum jerkum (fermented juice of plums) with your pub grub by the fire.

Or, you can bake red cabbage and black pear in cider vinegar. For how to cook this and black pears, courtesy of Worcestershire Orchards, see my Recipes section. Slow baking is needed as this is a very hard pear which never fully ripens. In medieval times this was an advantage, as if stored well it would be one of the few fruits still available in the new year.

It's so easy to forget this now, as we have all manner of fruits and vegetables shipped into our supermarkets out of season. That's not a very sustainable way of eating. In the past, as we had no choice, we had to eat preserved food. But, then we would surely have appreciated each new food as it came back into season.

Today, we think we have so much variety, but maybe in a way we have less; so perhaps it's time to really appreciate preserved food and seasons.

After consuming stewed and slowly-baked sweet delights, we can head off into Herefordshire. Here, as well as in the Vale of Evesham, you'll sometimes see sheep grazing in cider and perry-producing orchards. It's part of a sweeping band running down the western part of England. Somerset cider (often called scrumpy) is more well-known. But, we're now in the large cider-producing region of Worcestershire, Herefordshire and Gloucestershire.

Charlie Pye-Smith, on his journey through cider country in *A Land of Plenty: A Journey through the Fields & Food of Modern*

Britain[17], invites us to drink cider to help conserve cider apple orchards, particularly if you favour cider from independent brewers. He says that along with them, you'll be helping wildlife that depends on spring flowers, possibly the survival of heirloom varieties and storage of carbon in unploughed soils. He picks out a line from the famous *Cider With Rosie* by Laurie Lee.

Never to be forgotten, that first long secret drink of golden fire, juice of the valleys and of that time, wine of wild orchards, of russet summer, of plump red apples, and Rosie's burning cheeks...

These orchards haven't all disappeared. Let's support them.

Keep Going West for Hops

Fancy a beer? If you like beer, then go west in autumn to where an army of people have just harvested the hop fields, and now brewers are at work. They're flavouring and preserving ale with hops to make beer. It's the hops that make the beer, giving it a more bitter taste and preserving it for longer. From the Teme valley, west of Worcester, stretching out along the English/Welsh borders, there are fields full of hops scrambling up tall poles.

Locals have grown them here since the 1600s, but the heyday of hop growing was the 19th century. Walk along hop roads and imagine the seasonal hop pickers that in recent times flocked there from industrial areas like Birmingham, the Black Country and Wales to pick hops at harvest time.

---

[17] *A Land of Plenty: A Journey through the Fields & Food of Modern Britain* by Charlie Pye-Smith; Elliot and Thompson

Hop pickers of Worcestershire in 1917. Courtesy of Worcestershire Archive and Archaeology Service

Apparently, the terrain, soils and 'dull' maritime climate makes British hops unique. Yes, dull maritime climate. Living here, as I write this today, that figures. The rain is falling and the river is rising.

Farm shops around here all stock locally-produced cider, perry and beer from small producers, but otherwise, you'd need to shop in a fairly large supermarket to find them. If you live near a cider and perry-producing area, support your small producers.

They're often revitalising traditional methods of fermenting and production, and may well grow heirloom varieties of apple and pear in small orchards. Small orchards that are wildlife-friendly, offering a refuge for insects, worms, wild flowers, fungi, mosses, birds and bats.

Some farmers still let their pigs loose into the orchards, where they happily snaffle up fallen apples, along with grubs and insects. That's feeding pigs old-style.

Terroir, Apisoir and Artisan foods

Market gardens, orchards and hop yards may be a more recent imprint on the landscape, but they've flourished here. They also take advantage of an ideal local climate that affects their 'terroir'. We probably associate the word 'terroir' with the wine-growing regions of France. Most of us know that wine can have a distinctive taste (even if we feel we can't detect it) that comes from the soils in which the vines have been grown, the way they've been grown and harvested, the microclimate, the local fermentation process, and more.

The same applies to cheese, coffee, tea and honey, for instance. They say you can taste the pasture in the cheese, or taste the place in the food or drink.

Some time ago, I came across a similar word in a podcast called *Inventing a new word: Apisoir*[18] by Root Simple. Here, Erik Knutsun interviews Michael Alberty, a wine writer, about his attempts to get Wikipedia to recognise his new word 'apisoir' (relating to bees and honey).Wikipedia rejected the 'apisoir' page on the basis that the word isn't in the dictionary. But, new words are made all the time, Michael Alberty points out, and goes on to explain why different variations on the word terroir could have value. I think they would draw us in to understanding what affects the taste of food and, in doing so, come a little closer

---

[18] *Inventing a new Word – Apisoir* on the Root Simple podcast; https://www.roo tsimple.com/2018/01/115-inventing-a-new-word-apisoir

to a particular landscape and food-producing culture.

Michael explains how hives around Portland in Oregon produce distinctly different flavoured honey. The taste is especially affected by the flowers on which the bees feed. On a vineyard near Portland (Cameron Winery), bees from a hive near a patch of cardoon flowers produce a honey which has a subtle taste that makes you think of a metallic green, like jade, with a hint of artichoke flavour too. Inevitably, the nectar from the cardoon flowers (related to the globe artichoke) affects the taste of the honey. The honey from another hive in the vineyard, which sits near some buckwheat, has a distinctive, earthy, barnyard-like taste.

Some people have said that this concept of taste and terroir is pretentious and rather precious, but it is a real factor. Coining a term like 'apisoir' can help artisan producers carve out a niche for their product. It's usually more expensive to produce honey from a small apiary, and to compete with imported, mass-produced honey, adulterated with corn syrup, that we see so much of on shop shelves. For small beekeepers, being able to draw attention to their distinctive honey can make a difference to their sales.

Erik Knutsen, who is a natural beekeeper, adds that it also says something about how the honey has been produced and how the bees have been treated, and as Erik says, bees who go on vacation are somewhat controversial.

You may have heard that hives are moved across the United States to pollinate almond plantations in California. But, going on vacation is a euphemism because it's anything but benign. Holidaying bees dye in numbers, most likely poisoned by pesticides used in orchards, and exposure to new pests and viruses.

We might ask whether artisan food is pretentious and elitist, but is it pretentious to care about the welfare of bees?

Sure, to some extent, industrial food producers have appropriated the word 'artisan' to sell food, and it also raises the question of whether artisan food is affordable. There are issues with the artisan label being used merely as a marketing ploy, but let's not throw the baby out with the bathwater.

Buying even a little artisan food draws our attention to what it takes to produce food. Buying locally, at more or less farmgate prices, can mean it's more affordable. Seeing that most people will find someone locally producing honey, why not try honey?

Now we have a special word for this concept; but before the industrialisation and globalisation of food, all food must have had a terroir.

Is Better Food Affordable?

It could be easy to dismiss artisan, organic, free-range and high-welfare food as unaffordable. It may only seem so, though, when we have so many different ways to spend our money. We're also not taking into account how cheap food costs us all – financially, not just in damage to our health and the environment.

What do we spend our money on? Cars, TVs, mobile phones, and all manner of useful or entertaining stuff. Stuff that most of us wouldn't want to be without. We also (many of us) like to spend money on wardrobes stuffed with clothes we rarely wear and cupboards rammed full with more cleaning chemicals than we need.

The key to change is working out where our modern spending is useful and valuable, and where it's questionable. It's within our power to change focus.

The problem of plastic pollution is getting much more attention in the mainstream media than polluting food. Recently, Hugh Fearnley-Whittingstall and Anita Rani, during the British TV programme *War On Plastic*, asked the occupants of an average street in Bristol to empty out the contents of their bathrooms onto the street.

There was an astounding amount of shampoo bottles, toothpaste tubes, make-up and packs of wet wipes. They'd already covered how loose vegetables and fruit cost more to buy than those packaged in plastic.

Some of the street occupants said they don't buy loose food because it costs more, and that they can't afford it. It doesn't seem right that loose food costs more, particularly when plastic is such a problem. But, on the other hand, clearly, we feel we can consistently afford wet wipes, but spending a little extra on plastic-free loose vegetables will break the bank.

How Have Our Spending Habits Affected Farmers?

Charlie-Pye-Smith, in *Land of Plenty: A Journey through the Fields & Food of Modern Britain,* travelled the length and breadth of Britain, interviewing farmers about their lives and how they produce the food on which we all depend.

The price they get for their efforts crops up regularly. Andrew Burgess, a grower of organic vegetables on the Houghton Estate in Norfolk, gives us a telling account of how drastically our spending on food has changed.

When my parents were born in the early years of the twentieth century the average household spent around 50 percent of its disposable income on food. By the time I arrived in the 1950s, the food bill still accounted for around one-third of average incomes. Today the equivalent figure is 10%. So, for every pound the average family earns, just 10p goes on the most important thing in the world: food....

Our spending habits have affected not only farmers, but ourselves too - the consumer. Driving down prices generally drives down the quality of the food. Better food seems more expensive, but in many ways it isn't. We don't necessarily see how we're all spending too on counteracting the negative effects of industrially-produced food. If as many people as possible bought into that better food list, the costs to us all would diminish, and it may kick-start more system-wide changes that would make better food available to all.

Regardless of whether we buy cheap, mass-produced food or not, we're spending more on food than we realise. In most countries, we pay for:

Water companies and government agencies to:

- Remove polluting agricultural chemicals from our water supply
- Subsidise industrial farming that uses fossil fuel-based chemical fertilisers on both cornfields and pastures, herbicides on crops, and pesticides or drugs to combat disease in crops or animals
- To police run-off of agricultural chemicals into our waterways
- To police animal welfare
- To monitor wildlife on farmed land
- To manage flood prevention

All these issues are affected by the way that land is farmed. That's before we even get on to how much we're paying into the

NHS, in Britain, or into medical health insurance premiums in other countries, for illnesses caused by plentiful but nutrient-poor food.

What if we spent less on taxes to pay to produce industriall y-produced food, to police agricultural pollution, less to water companies to filter out pollution, and more on good food?

Of course, low-income families are less likely to see the change as they're paying less or nothing in income tax. The key must be for society to change so that good, local food is affordable. We may pay more at the till, but wouldn't be paying into government subsidies, and taxes to cover pollution control and medical care for illnesses caused by poor quality food.

Take Food Back Into Our Own Hands

Our food is increasingly controlled by global corporations and big institutions, but there's a whole host of people joining together to disrupt and regain control. They're part of the food sovereignty movement. Worldwide, they're people who want access to land to grow food, farmers who want to grow genuinely sustainable food and get a fair price for it, skilled traditional food processors, fair trade organisations and ordinary consumers.

Farmers, who over the last few decades embraced 'Big Ag', are returning to seed-saving and distributing seed, instead of bankrupting themselves buying patented hybrid or GMO seeds; they're managing pests using biodiversity, not pesticides; managing weeds using people power, not herbicides; selling their produce direct to consumers. Overall, those supporting the food sovereignty movement are for a locally distinctive diet, not an homogenous diet. They're for food high in nutrients, and more natural food processing, rather than modern processing that destroys nutrients.

Relatively, artisan, organic, free-range, high welfare food may seem more expensive (when you ignore the hidden costs of cheap food), but it's better for us and the environment. **On the other hand cheap food will erode local distinctiveness and cost us the earth.**

What Can We Do?

You may well feel that there's little you can do until governments influence how we produce and sell food. But, the zero waste and green living movements believe that we can all play our part now, today – even if it's just a small introduction of more locally-pr oduced/artisan/wildlife-friendly food into your kitchens.

You could:

Review Your Spending

Ask yourself, how much are you spending on food, and how much on goods you don't need? Could you ditch expensive wet wipes and a cupboard full of cleaning chemicals for **better food**, and a simple, natural cleaning regime?

This might not suit everyone, but in our household, we've gone back to paying for food in cash, to a budget. We put the cash in a spare wallet, which we take out food shopping with us. If we forget the wallet, we adjust the money in it when we get home. Any extra unspent cash goes into next week's budget. It keeps us focused on what we're spending.

Waste Less Food

Why buy food only to waste it? Use your leftovers. I won't cover this much here as plenty has been written about this, other than to say that old favourites like cottage pie and bread and butter pudding are ways to use up leftovers.

Other ways are:

- Using the fat from meat
- Making stock from bones, poultry carcasses and past-their-best vegetables
- Nose-to-tail eating
- Using stale bread to make breadcrumbs, croutons and toast
- Cook more from scratch with a pared-down range of ingredients, which makes it easier to keep track of your food stores

Grow Your Own

Depending on your inclinations, this might be first on your list. You could, though, hardly provide yourself with cheaper and more nutritious food than that which you have grown (or reared) yourself.

If you have little space, there's much to be said for the small productive garden. Imagine going up in a balloon over where you live and seeing tomatoes, beans, peas and green leaves growing in every back garden. And, on balconies and roof gardens. A few chickens strutting around providing fresh eggs wouldn't go amiss. Wouldn't that be a good sight?

If you don't have a garden, if you're in the UK you could rent an allotment (a plot of land to grow vegetables or rear chickens, for instance). Some people even take over patches of unloved waste ground for growing food by guerrilla gardening. You might even have a neighbour with a garden they can't look after. They might be happy for you to grow vegetables there, and share some of the produce.

Walk, Cycle, Take the Bus

If you can, make shopping take more effort, and get fit in the process. If you have to carry your shopping you can't buy too much at once, and you have to be more conscious about what you buy. Getting the bus makes you do this, even if getting fit isn't much of an extra gain.

This is what I've been doing lately - cycling to a farm shop over the river. It takes longer to shop, but I'm getting some much-needed exercise, and I have to think about what I buy. Otherwise, I might not get my shopping into the bike panniers. Then how would I get it home?

Quick Takeaways

- Ask what's worth spending money on, compared to food
- Shop with cash?
- Clear your cupboards
- Grow some food, indoors or outside, on your, or borrowed, land
- Make food shopping take more effort

* * *

We've just toured round a diverse landscape producing diverse food, where there's no one single method of producing it. There are farms growing on an industrial scale. The small family farm is still there, though, producing food using age-old ways. This is a real landscape producing real food.

What about those statistics though, and those greenhouse gases that we can't see?

6

Whose Facts Are Right?

After reading the preceding chapters, you may think it's all very well painting a charming picture of bygone ways of farming and eating that, in the past, worked with the local landscape. But, we live in a world of rising population (not to mention climate change), and there's all that methane to deal with. Yes, methane has hit the news big time in the past few years.

Farming cattle and sheep in the uplands, and on poorer soils of the lowlands, has always made sense. Along with nutrient-dense food, their manure fed the soil on which we grew grain, fruit and vegetables. And, in some places, poop fertilisation of soils still reigns. But LIVESTOCK PRODUCE METHANE! We keep hearing this message.

So, all that I've written is for nought. Or, is it? It depends on whose facts you think are right. Facts are not truths. They're merely interpretations of data. And there are countless ways in which we can arrange the data before we get our calculators out and interpret - if we're dealing with numerical data.

If you've followed some of the arguments over what is considered to be sustainable food in the media, or have only come across the headlines, are you bemused? I certainly have been. It can all seem very contradictory and confusing.

Why I Became Suspicious of Popular 'Facts'

At work, I spend a fair amount of time at a microscope doing routine work and, whilst I do, I listen to podcasts. Three of my favourites are BBC Radio 4 podcasts: *The Food Programme*, *Farming Today*, and *Costing the Earth*.

Four years ago, the topic of sustainable food started cropping up on these podcasts, and there was much said about emissions from livestock and efficiency of land use for producing food. Emissions were something I'd never really thought about before. But where producing food is concerned, efficiency of land use is a topic on which I've spent my whole working life.

I've mentioned that I work in archaeology, and it just so happens that I've spent the lion's share of my time looking at agricultural and kitchen waste.

Here I am, at the microscope, screening dried debris from soil samples from various excavations. Most contain burnt residues from cereal processing. Next to me are bags of animal bone I'm waiting to send to a specialist.

I pick up a clutch of bags from a site in north Warwickshire, while one podcast guest tells me about the methane belched from the bellies of cows, and how problematic it is.

I see that burnt crop processing waste in these samples is typically sparse, fitting the pattern that we know for the area. The pattern was generally of mixed farming, but, leaning more towards pastoral. The samples are either Bronze Age or medieval in date as two settlements of different date lie in the same fields. The results are the same, for in this location the balance of pastoral and arable farming has stayed broadly the same for thousands of years.

The thought crossed my mind that these farmers had no idea that their cows were emitting so much methane. How could they have known? I shrugged my shoulders. After all, there weren't so many people and cows (or sheep) then as there are now. And, they weren't burning fossil fuels, which also produce methane.

There were different guests on these programmes with different messages to convey. Some of what they said resonated with me, and some didn't. The next message that came across was about how inefficient it is to farm animals. And how it wastes energy and resources.

It stopped me in my tracks. I had a little pile of bags of debris beside me at the microscope, all from the area of the Midlands that has always relied on mixed farming. Livestock and animals together as a strategy for keeping up soil fertility on less than perfectly fertile soils. And, for producing nutrient-dense food in a world where the food you ate came from the ground beneath your feet, and had to respect growing seasons.

Mixed farming was, and still is, surely the most zero waste way to produce food? What on earth was going on? Who entertained such ideas?

A voice of reason, in one particular podcast, came across. That was the voice of Simon Fairlie who related a story about Bramley, his cow. For he was moved to work out how Bramley could have possibly consumed the vast quantities of water in his lifetime that squared with current statistics circulating in the popular media.

I asked myself whether the statistics, and ideas that puzzled me, related to intensive livestock farming? I followed up by reading Simon Fairlie's book - *Meat: A Benign extravagance*[19]. He has some interesting points to make about the water footprint of beef.

Bramley the Cow and his Unfathomably Large Water Intake

How to fit a quart of water into a pint-sized cow? It's a question Simon Fairlie poses as a subtitle to a chapter in his book called *Hard to Swallow*.

He was bemused by a statistic he'd seen circulating in the media at the time of writing. George Monbiot of *The Guardian* newspaper had used it, as had countless other journalists. He felt compelled to chase it down.

He'd read that we need to use 100,000 litres of water to produce a kilogram of beef. It was becoming a popular statistic, particularly with those who believe we must transition to more plant-based eating.

Fairle worked out that's equivalent to an acre an inch deep in water. Even more mind boggling was the thought of what an entire Bramley could have consumed. Bramley was only a small Angus/Jersey cross steer (male). At slaughter, he would have consumed an acre 10 foot deep in water.

He says of Bramley:

How he managed to achieve this feat I am at a loss to explain, since all Bramley did while he was alive, was hang out in a field and eat grass, without ever manifesting any unusual appetite for water.

---

[19] *Meat: A Benign Extravagance by Simon Fairlie*, Permanent Publications (1st Edition)

Fairlie then ponders the question of where the water could have come from in Bramley's life, and where did it go? All he did was eat grass. Some of it exited his body in urine and manure straight back onto the field. All to the good of the field, of course.

Some of it transpired on his breath, travelled somewhere on the wind, then it made it back to earth as rain droplets. You could include the water in his mother's milk. Either way, it was hard to make Bramley's water consumption match the statistic he'd seen.

So, where did the water come from in the researcher's world? He traced the statistic back to figures produced by agronomist researchers David Pimentell and Robert Goodland.

The obvious answer would seem to be that this statistic relates to industrially-farmed cattle in US-style feedlots where we confine them and feed them on food from irrigated cropland. But, to his surprise, the majority of the water accounted for came from hay - a traditional winter feed for livestock.

How could hay from a haymeadow take so much water to grow? After all, hay meadows grow every year with no extra applied water. I like the erudite comment:

Over large parts of Britain, it is difficult to stop hay growing, which is why people use lawnmowers.

Fairlie points out that half an hour on a search engine is enough to dig up a bewildering array of estimates of the amount of water needed to produce a kilo of beef. Most of us would never have the patience to chase up and analyse this data. I certainly wouldn't, as maths was never my strong point. So why worry?

The problems start when one outlandish statistic (quite possibly wrong) makes its way into mainstream journalism, and journalists repeat it over, and over again. We, mere mortals, soak up the information and then suspect foods of being 'bad' for reasons we needn't.

Methane and Cow Belches

We hear so much about methane, and most of the negative facts relate to livestock and meat consumption. Methane has been touted as a most potent greenhouse gas. Ruminant livestock (the grass eaters - famously, cows) have really taken the stick for belching out huge quantities of the gas. Poor cows. We keep hearing it - we judge them GUILTY.

What is rarely mentioned is that most of the popular statistics still circulating go back to one report. It's called *Livestock's Long Shadow*[20]. It was produced in 2006 by the Food and Agriculture Organisation (FAO) of the United Nations.

Many writers have criticised this report for good reasons. But, their criticisms are squashed by the might of enthusiasm for its message when it was first published. That's enthusiasm by journalists, various food interest groups, and green-living influencers who follow in their wake.

The main message in the report is that it's best to raise animals in confinement, where they can be brought to maturity quickly by feeding them with specialised feed. The authors quote statistics on the amount of methane livestock (like cattle) produce, the effluent produced (manure and urine) and their water demands.

Where methane is concerned, the authors use statistics on

---

[20] *Livestock's long shadow: environmental issues and options*; http://www.fao.or g/3/a-a0701e.pdf

how much cows belch out, for it is mainly their mouths and not their rear ends that emit the offending gas. However, they fail to consider all the methane emitted when we turn oil or gas into agricultural chemicals, such as fertilizers, herbicides and pesticides. Why? Because they were using statistics produced by the United Nations. And, the UN categorise these emissions under 'industry'.

This omission is not the only one. Methane is made up of carbon and hydrogen. But, nowhere in the report is the cycling of carbon in methane around the natural environment taken into account.

In short, carbon makes a journey through the air, and into the plant, which the cow eats. The plant (let's say grass) takes in the potent greenhouse gas, carbon dioxide (CO_2), which it processes to gain energy and grow. At the same time, it emits planet-healing oxygen. The cow eats said grass, swipes some carbon (and oxygen) for itself. In the process, microbes in its gut produce methane. Out it comes, and hence all the trouble with methane.

What many have pointed out, though, is that carbon in that cycle is just cycling around the natural environment. Poop fertilises the soil, releasing carbon in organic matter back into the soil. No new carbon is being produced as long as the pasture hasn't been treated with agricultural fertilizers and is managed well. But methane is a far more potent greenhouse gas than carbon dioxide, we hear.

A dangerous gas, some would say. Others point out that methane only has a lifespan of around 9 to 15 years. Carbon dioxide, on the other hand, stays in the atmosphere for a couple of hundred years.

Put it this way. Methane from all the sheep, pigs and occasional cows that lived on Red Rock Farm, where my husband grew up, from the mid-1970s through to early 1990s has long-since disappeared. It will have been broken down into carbon dioxide and water, but then taken up again by the pastureland all around. The carbon dioxide, nevertheless, released during every car journey you have taken, the chemicals applied to crop lands, in heating your home by gas and electric, in your lifetime has come from fossil fuels. That's carbon that has long been trapped underground and now released into the atmosphere.

Staggering. And, we blame cows.

If you're bamboozled by all this, I'm sure you're not alone. I suspect that one day there will be a report going around called *The Long Shadow of Livestock's Long Shadow* - once we've realised the problems of getting too sidetracked by a single study. There are problems relating to using crops like soya (mainly the crop residue) to feed livestock, but that issue isn't as simple as it may seem either.

Can we Feed the World?

That's a really big question. It's probably one which has been on the back of your mind while reading this book, and might make you doubt that we can leave modern industrial farming methods behind and return to more natural ways of farming.

We predict global human population will reach 10 billion by 2050. How can we possibly feed all these people? I've seen claims that we can feed all these people, and even up to 16 billion people. I don't know where these estimates come from, but they seem to be based on the knowledge (backed up by much research) that small-scale farmers feed the world, and that small farms are more productive.

A figure repeated many times has been that small-scale farmers produce around 70% of the world's food on less than 30% of the land. This leaves the conclusion that large-scale industrial agriculture only produces 30% of the world's food on around 70% of the world's arable land. Not only are small-scale farms more productive, but women working on these farms feed the world. They may rarely be landholders, but they are land workers and market sellers.

This is low-input farming, working closely with nature – often using age-old knowledge and techniques. In developed countries, mainly, we can add the benefit of modern technology to the tried and tested techniques, where it is beneficial. In time, this will spread more. Farmers use the internet to get information they need and to reach customers. I'm no expert, but small-scale digital technology seems to be helpful in many ways without causing the damage that we've seen from agricultural chemicals, massive machinery and huge monocultures.

Does this mean you need to go wholesale into either becoming a homesteader, or buying all your food from small producers of local, organic, free-range, high-welfare, artisan food? I can't give an answer, except to say that if many of us integrated just some of this food into our lives, we would make a big difference.

Wanting to do it, and seeing the reasons to do it, make a big difference in achieving it. Once you start, it's likely to become habit-forming.

What Can We Do?

What if you don't want to embark on a PhD in gaseous chemistry and ecology in order to understand it all? My thoughts are, reconnecting with where our food comes from is a better place to start. And, if you can make real-world connections with the people who grow and rear food, or grow your own, even more so.

Even if you live in a city or a landscape transformed by industrial agriculture, there are still ways to connect with people who grow food, so that what's on your plate has more meaning. Finding out more about it doesn't have to feel like going back to school.

If you really think, though, that you must get a better grip on the statistics relating to the carbon footprint of food, then you could:

Dig Beneath the Headlines

If global statistics on what makes sustainable food matter to us, we should ask ourselves: 'Who did the number crunching?' 'What do those people know about farming?' They may sit in university and government offices and have a very different world view to those who pull on their wellies every day and open a farm gate to tend their cows or check on their wheat crop.

This isn't to say that all statistics are wrong, but we can go wrong if they are our only guide. And, if we don't understand where the raw data comes from and how the numbers have been 'crunched', how can we trust the message?

Even data and conclusions coming out of the most respected institutions in the world, (the United Nations and Ivy League universities, for instance) are being questioned. Most of us would be hard pushed to find the time to dig deep into this world of statistics and do that questioning. So, the real, tangible world around us is not a bad place to start, alongside keeping an eye out (or an ear to the ground) for alternative views.

The pace of science is slow. Science works by testing a restricted number of variables at any one time. And, studies on greenhouse gas emissions, or the carbon footprint of food, are few and far between compared to studies relating to climate change. We are nowhere near a consensus. It's a gradually evolving, messy sphere of numbers and different voices; from the number crunchers, to the big corporate food producers, and those with mud and manure on their wellies.

If we're following news on what is today's sustainable food, whose words are we reading? What do the writers know? It might not occur to us to ask these questions. Someone writing for our favourite newspaper might have deep knowledge of farming, or have never produced a carrot or a cow in their lifetime. Of course, they don't have to get their feet muddy to have reliable opinions on this. It all depends on how much they've researched and talked to people at the coalface of producing food.

If you're reading online, it doesn't take much to, every now and again, click on an author bio to get a handle on who they are and what they know.

Follow the Money

Money, money, money..... we can't leave money out of this. Academic research has to be paid for. The funding may come entirely from an independent academic research fund, or increasingly, partly from industry. The financial interest of industries funding research can have a bearing on the conclusions of the research. It's been known for decades to have been the case.

Typical industries which have an interest in funding this research are: big food companies like Unilever, Nestle, Cargill,

Danone, Kelloggs; multinationals like Google; and agricultural chemical giants like Syngenta, Bayer and Monsanto. These companies also donate large sums of money to well-known newspapers.

Money talks, they say, and it's highly likely to be grabbing the headlines and influencing in all sort of ways we've not thought of. Tune your antennae to the possibility and they will surely twitch.

Quick Takeaways:

- Ask, do statistics matter to you? Or would the real world outside your door be a better place to start?
- If statistics matter to you, look for opinion pieces that give more detail about where the data comes from and who did the number crunching
- Check out all opinions, not just popular opinions
- Check out who funded what

* * *

Facts matter, but I think food should be about more than just numbers without context. About emissions and methane. Emissions matter, as does the life and death of animals we farm, and the wildlife that depends upon that land. But we need to see the big picture, and once we do, we have a better guide to our choices. There's much you can do......

7

What Can We Do?

How can I live better? That's what many people are asking. When it comes to eating better food (better for us and the planet), we want to know what we can do, not just what governments and international agencies can do.

I don't believe there's one 'fix', but we needn't be rudderless and confused. This book has focused on *one* good starting place. And that's 'place', for there's **power in place**. Think #PlaceBasedDiet or #PlaceBasedFood, regardless of what you hear on social media. We're so used to hearing the words 'plant-based diet' - a concept that seems to have turned into our global one-fix solution for protecting us from global warming and for saving animals.

There's no one fix that applies to everyone, everywhere. If you start with the ground beneath your feet, however, then you train your mind on how food is ultimately very connected to landscape, soil, water and climate. You cease to view food as a single tick-box in a list of to-do's if you want to be a more sustainable, conscious or ethical consumer.

1. Start with Place

The word *place* forces you to think beyond a single homogenous fix; less like a consumer in the middle of a supermarket and more like a farmer or producer matching crops and animals to good soils, to poor rocky ground, wet, low-lying plains and local climate. You might think like a forager, or even hunter. If you're disconnected, then it entices you to reconnect and enter a new world.

For other issues, such as reducing plastic pollution, the message that's been helping the most is to take one step at a time, or, start from one place. You can take the same approach when it comes to food.

If you eat like your ancestors, you'd eat more from the ground beneath your feet. Or as mentioned in the introduction, from the water lapping up against banksides and shores.

Even if your recent ancestors came from somewhere very different to where you live now, you can always adapt the traditions of your ancestors to the place you live now. After all, throughout human history, entire communities have migrated and done this.

I believe that starting here will inspire and inform you. Even bring a breath of fresh air into the issue, because you'll need to get out into the fresh air.

You don't need to join a diet tribe to do this, although there's the term *locavore* that might apply. Someone, somewhere, will tell you that a particular way of eating is THE way to eat. If I appear to be doing the same, it's not one way of eating regardless of where you are in the world.

2. Ditch the Diet Silo

We've hunkered down in our silos. We're omnivores, primal, vegetarian, flexitarian, reducarian or vegan, or veggan (a new word meaning a vegan who eats eggs). We've subscribed to the group, we wear the badge and the school uniform.

We've battened down the hatches. And, quite probably, we don't like talking to each other. It's like someone subscribing to a different diet is of a different religion.

There is, though, something we all have in common. We all live somewhere that produces food – nutritious food. Excellent food is out there that sustains environment and wildlife. Yet, the food movements and diet silos we've become familiar with have come about because we're disconnected. They don't have any roots or ties to a particular landscape or culture, like many cuisines around the world. So how do we find our roots?

3. Reconnect with Food

Look around you. Notice the countryside. Even if you live in a city, there's still countryside and farmland around the outskirts.

Most of us live in cities and do a supermarket sweep once a week or month, grabbing food packaged in plastic. A cauliflower in a plastic wrapper has become divorced from the soil on which it was grown, and the farmer who grew it.

But need it be like that? Do we have to remain so disconnected? Some time ago, I came across podcasts by Harold Thornbro of *The Modern Homesteading Podcast*[21]. There's a prompt about

---

[21] *Modern Homesteading Podcast*; https://www.haroldthornbro.com/blog/cate gories/podcast

that disconnect in the podcast intro, followed by one of the solutions:

There's music, then someone says.... '*We have allowed ourselves to become so disconnected and ignorant about something that is as intimate as the food that we eat*' [cows moo] than another voice says '*...be prepared to grow your own for victory* [the Grow for Victory campaign in the US during the Second World War]'.

Play these podcasts and you'll hear it in the introduction. The answer is, as Harold suggests, there is much you can do from your small-town location. You don't need a PhD in food and farming, and you don't need to be a farmer in the country.

The best way is to grow or rear your own at home if you can. Even if it's only a little I wouldn't underestimate its value. The Second World War, dreadful though it was, kick-started a home-producing revolution.

We produced an astounding amount of food in our gardens, balconies, parks, and on allotments. This happened because of the Dig for Victory and Victory Garden campaigns in good ol' Blighty and in the US.

Imagine going on a balloon ride over where you live and seeing food in every garden and back yard. Even if these are just one or two common vegetable crops like tomatoes and runner beans. Every tomato bush or vine takes pressure off farmland to produce in unsustainable ways.

Think of the reduction in ground covered by glass houses, of the reduction in use of pesticides and herbicides. Think of the water saved (if you're using rainwater). For each household, it wouldn't take much, but we'd bring back productivity to front gardens, back gardens, patios and balconies.

4. Ask 'Where Does My Food Come From?'

What if most of your food could come from only a short distance away, using little fossil fuel to get it through your front door? It sure could. If there's something about your eating habits that sway you towards a lot of high mileage food or unknown food mileage, you can change that scenario.

As well as, ideally, producing some of your own food, you may need to change your shopping habits.

When it comes to food you can't produce yourself, if you think differently, you need not plunder the global food bowl. All types of food (meat, dairy, eggs, vegetables and fruit) traverse the globe, refrigerated, in huge tankers or on airlines. Yet most of this food is available on your doorstep too.

And, if you learn different ways to prepare and cook with it, it's bound to become more interesting and sustaining. A different cooking technique, some herbs and spices - that's what traditionally we've used to vary our food. Romans introduced most of our well-known herbs and we imported our spices, so we've used outside influences and imports for a long time, but today, where importing food is concerned, the sky seems to be the limit.

We all like some imported food or drink to add to what we have on our doorsteps. We've been trading food and drink produce for thousands of years, but never on this scale. That's what we're grappling with - finding the right balance.

5. Discover the World Outside the Supermarket

If at all possible, get out of the supermarket. There is a world outside of them (if you go looking). Although it probably won't mean you never go in one again. After all, you can buy organic, free-range, higher welfare and sometimes local food in a supermarket. The world outside will broaden your perspective though. Supermarkets are very artificial environments that can disconnect you with food.

Not everyone has easy alternative options accessible to them. But, the more you change your mindset about where your food comes from, the more you can see ways to change for the better.

For a PlaceBasedDiet, I think of Farm to Table, though you may think just as much Sea to Scullery, depending on where you live. Here are some places out there:

- Farm shops (bricks-and-mortar and online)
- Urban farmers' markets
- Independent grocers, butchers, fishmongers and cheese-mongers
- Individual smallholders online
- Community Supported Agriculture (CSA) farms

Aside from producing food yourself, all these ways of buying food help you support farmers or fisherfolk who look after the environment, their animals, their soils or the sea, whilst contributing to the local economy. They're often farming or fishing in a traditional way.

If you live somewhere where the local food-producing pattern has broadly stayed the same for thousands of years, it's probably because that's the most sustainable way to produce

food. This is the case for large areas of the British Isles where I live.

6. Bump Up Your Knowledge of How We Produce Food

As you're reading this, you must be interested, and the best thing you can do is harness your interest. The previous chapter shows us how confusing and bamboozling information out there can be. A good place to start is with people who produce their own food.

I've come across lots of homesteaders out there on social media. If you read their blogs, they all have their own stories about how they came to do this. Most weren't born into it. They didn't necessarily have much money either.

Hear from Gently Sustainable about '*How to Start a Homestead With No Money*'[22]. Some dived in at the deep end. Some planned and researched first. They nearly all talk about a steep learning curve; the trials and tribulations.

I'm also a self-confessed armchair farmer. There are so many fascinating books written by farmers and smallholders for anyone to read. They want us to know about farming. These are people who care about what they do. They farm closely with nature and care about the animals they keep. The big agri-industry farmer doesn't tend to write books for the general public. There may be a reason for that.

Most of the books I've read are British: you may need to search for those more relevant to where you live. One I've read, *Letters*

---

[22] *How to Homestead With No Money* by Gently Sustainable; https://gentlysusta inable.com/how-to-homestead-with-no-money

to a Young Farmer[23] I can't praise highly enough - it's American.

If you're really getting into PlaceBasedDiet, you could think more about how you might produce more of your own food, even if you don't now. Call it the 'Imaginary Homestead'. This could be because you're thinking of starting up your own homestead or smallholding, or it's more curiosity about the how.

Most of us are first world consumers. We've never had to rely on producing our own food. Big influencers are often people with no experience of this either. And, yet, we rely on what they say, almost with religious zeal, it seems. What if you imagined how you would produce your own food, given a small plot of land? You would have to think about terrain, soils, local climate and much more.

There are ways in which thinking like a homesteader could be illuminating. We might even change the way we eat.

* * *

Do We Need Unprecedented Change?

In first-world cultures, we have the right to individual choices. But we'd be in a far better place if a sense of the balance of nature were to guide us. Then the farming landscape need not be in trouble. Let's celebrate and learn from local distinctiveness, farming heritage, and time-honoured skills honed over generations.

In a modern world, troubled with the threat of climate change

---

[23] *Letters To a Young Farmer: On Food, Farming and Our Future*; http://www.letterstoayoungfarmer.org

and a rising population, we might question whether age-old ways of farming heritage are relevant.

Our ancestors may have turned their toes up and are six foot under, but let's not forget: they had knowledge that has stood the test of time. Knowledge we've pushed out into the margins of society, but ought to come back to centre-stage.

The 2018 the IPCC Special Report on Global warming of 1.5°C[24] states that limiting global warming to 1.5°C would require rapid, far-reaching and unprecedented changes in all aspects of society. We see the words 'unprecedented change' bound up with much of what we read in the media about food in the future. The change seems to be synonymous with eating in ways we never have before. There are some, though, who are saying we need to eat in ways we always **used to** before.

* * *

I've mentioned the poem, *All bread* [25] by Margaret Atwood which shows us that wheat should be fed from the decay of small mammals, insects, animal dung; all melting into the soil, sucked up through the stems, the grain and into loaves of bread. It's a messy business, but a natural one.

What would you prefer? The bread made from wheat fed with cow dung, moss and bodies of dead animals? The rotting matter

---

[24] *IPCC Special Report on Global warming of 1.5°C*; https://www.ipcc.ch/sr15

[25] *All Bread*, from SELECTED POEMS II, 1976-1986 Copyright © 1987 by Margaret Atwood. Reprinted by permission of Houghton Mifflin Harcourt Publishing, Oxford University Press, Canada and Curtis Brown

living in a field that is at least close to a natural ecosystem? It might sound undesirable, but that's the basis of good nutritious food and, for instance, bread that has a distinct, nutty flavour to it. It's where wildlife thrives.

Or, would you prefer the bread made from wheat fed with coal and oil (fossil fuels) in the form of chemical fertilisers, pesticides and herbicides? Bread from wheat grown in a place where few living creatures thrive. A place devoid of animal dung, on which dung beetles thrive, and all manner of creatures connected in a big food web.

I know what I'd prefer.

8

Recipes

These are recipes for a selection of the food I've mentioned in the book – food that fits with the local food culture of the area we've just travelled around. You could also find food that makes the best use of local food produced around you.

A Word About Fat

Ideally, when cooking meat, cook with fat from the same type of meat. Save the fat after roasting meat to make dripping. To do this, tilt the roasting tin or pot, so that the fat settles on top of the meat juices. Skim off the fat, avoiding the meat juices as much as possible. However, if you are regularly using the fat, small amounts of meat juices won't cause the dripping to spoil, as you'll use it before spoiling can happen.

A Word About My Approach to Weights and Measurements

I'd rather not give you any weights and measurements, and where I can do without, I leave them out. That's because I suspect that sometimes the thought of concentrating on a list of specific quantities can put us off trying different food when we're busy. At least, it works that way for me. And, when you want to make more or less than the recipe dictates, you end up multiplying and dividing figures, when judging by eye often works very well.

I've listed imperial and metric weights and volumes.

First, I offer two meat recipes characteristic of hilly country, where I started this journey, and we then move on to food that suits the following chapters.

PIES

Spiced Mutton Pie

Spicy mutton or lamb probably doesn't sound too odd to most people, as we're most likely to think of a spiced meaty tagine recipe or a curry. Spicy, fruity meat might sound odd, but in the West Midlands, fruity meat dishes are traditional. We're in orchard country, where farmers and smallholders often grazed sheep and pigs in orchards. Their grazing on fallen fruit, along with grass and herbs, affected the taste of the meat. Perhaps this is why we traditionally accompanied meat with similar tasting herbs, sauces and ingredients?

A spicy, fruity mutton pie is a frugal, old English classic which makes use of cold meat left over from a roasted joint. Find a size of pie dish that would work for the quantity of meat you have.

For the filling:
    Cold mutton (or substitute with lamb)
    Cooking apples, chopped
    Lard, ideally from mutton or lamb
    Sugar to taste
    Ground nutmeg
    A few prunes or sultanas

Method:

- Make a shortcrust pastry lining (see below) to fit your pie dish, and leave to cool.
- Use equal quantities of cold, cooked mutton (include some fatty pieces) and coarsely chopped cooking apples to fill the pastry case. Fill the pastry case with alternate layers of mutton and apple with a light sprinkling of sugar and ground nutmeg on each layer.
- Include a few prunes or sultanas in the mix.
- Add some mutton fat, dripping or lard on the top and add a pastry top. Bake in the oven at 200 - 220°C or 400° - 425° F until the pastry crust is lightly golden.

Short Crust Pastry:
    8oz/225g/1 1/2 cups plain flour
    2oz/60g/4 level tbsp butter
    2oz/60g/4 level tbsp lard
    Approx 2 tbsp water

A batch of 8oz pastry is sufficient for a standard 8-inch (1

1/2 -inch deep) pie or flan dish. Adjust quantities up or down based using, half fat to flour and sufficient water to bind the ingredients together.

Method:

- · Mix the flour and salt in a basin, and rub in the fat
- · Using a knife, mix in the cold water to form a stiff dough
- · Leave the pastry to rest in the fridge or a cool place for around 30 mins
- · Turn the dough out onto a floured surface and roll out

Tip: keep your hands cool while rubbing the fat into the flour, and while working the pastry.

Fidgit Pie

This is a pie which is particularly associated with Shropshire – although I don't know the history behind it. I probably could have bought a ready-made fidgit pie in the Shropshire town of Church Stretton (where we started our journey in this book), but it was a Sunday, and the small, independent shops were closed. If not, I could go to Ludlow Food Centre nearby and pick one up. Meat, alcohol and fruit go well here. So, if you want to make your own, here are some generic instructions.

You could make this as a single pie, or small individual pies. For shortcrust pastry, see the Spicy Mutton Pie recipe.

For the topping:
   Potatoes (4 medium-sized potatoes for an 8-inch pie dish)

Wholegrain mustard mixed with honey (around 1 tbsp)
Large knob of butter
Cream or milk (enough to cream the potatoes)
Grated cheddar cheese, or similar

Method:

- Make the shortcrust pastry and leave to cool (see Spicy Mutton Pie recipe)
- Use equal quantities of gammon and coarsely chopped cooking apples, enough to fill the pastry case
- Peel and cook potatoes for 20 mins, ready to make mashed potato
- Lightly fry the gammon chunks to seal the meat. Cover with cider and simmer until the gammon is soft and has soaked up the cider
- Add the chopped apple chunks and toss in the pan, with the gammon, for a couple of minutes to glaze the apple chunks
- Drain the potatoes and mash with butter, milk or cream and mustard and honey mix to taste
- Line a flan dish, or similar, with the pastry. Add the gammon and apple filling, top with a layer of grated cheese, and mash potato
- Bake in an oven for around 20 mins

PANCAKES AND BISCUITS

Staffordshire Oatcakes

These were staple food in the potteries of Stoke-on-Trent and much of Staffordshire. You could make the oakcake mix and cook within a couple of hours, or pre-soak your flour (see below) in order to break down anti-nutrients in the flour, as mentioned in Chapter 4. However, remember that if you eat plenty of nutrient-dense food (like meat, milk, cheese and eggs), it's not essential to do this. It's more important if you're very dependent on grains, and other foods like pulses, nuts and seeds in your diet.

Ingredients:
    5oz/140g/1 cup oatmeal
    2oz/60g/ 1/2 cup wholemeal flour
    2oz/60g/ 1/2 cup strong white flour
    1/2 tsp salt
    1/8 tsp dried yeast
    3/4 pint warm milk and water

Method:

- Mix the oatmeal and flours in a basin. Make a well in the centre, and add a little of the warm liquid. When this starts to froth, work in the rest of the liquid and salt to make a batter. Leave in a warm place for one to two hours, or until you see a few bubbles in the batter
- To cook, ladle some batter into a frying pan (with a little fat to coat the pan) and fry as you would to make pancakes

Staffordshire Oatcakes (pre-soaked flour version)

Mix the oatmeal and flours. Prepare the ferment, as above (ideally, with buttermilk/kefir and water) but leave out the salt. Leave to rest in a warm place overnight, or insulate the bowl with towels or a blanket. In the morning, work in the salt, and cook as in the recipe above.

Sweet Oatmeal Biscuits

This recipe is based one from my great grandmother's cookbook: *Coombs Unrivalled Cookery for the Middle Classes* by Miss H H Tuxford, with my own additions for metric weights or tablespoon measures.

Ingredients:
    7oz/200g/1 cup medium oatmeal (generous cup)
    3oz/85g/ 6 level tbsp butter
    5oz/150g/1 cup plain flour
    5oz/150g/3/4 cup sugar
    1/2 tsp baking powder
    1 tbsp milk
    1 egg – beaten

Method:

- Mix dry ingredients (oatmeal, flour and sugar) in a bowl, rub in the butter and add the baking powder
- Mix in enough beaten egg and milk to bind into a stiff paste
- Roll out thinly and cut with a plain biscuit cutter into round.
- Place on a buttered or oiled baking sheet and bake slowly at 180°C/350°F for 15 minutes, or until golden brown on top

Soaked Spelt Cracker

Biscuits made from soaked flour usually use buttermilk or kefir.

This is my adaption, which I've trialled and found it produces a good biscuit:

Ingredients:
    8oz/225g/2 cups wholemeal spelt flour
    1/2 oz/15g/1 level tbsp butter
    4 fl oz/1/2 cup buttermilk/kefir
    2 tsp baking powder
    A little water to make dough pliable

Method:

- Add the buttermilk or kefir to the flour in a bowl, to make a mix of 'crumble' consistency and leave overnight in the fridge or a cool place.
- In the morning, work in the baking powder and the butter. Mix into a stiff dough, adding a little extra water, and bake at 220°C/428°C for 10 mins.

FROM THE ARCHIVES

The following two recipes are reproduced from *A Slice of the Past: A selection of recipes from Worcestershire Archives*, with kind permission of Worcestershire Archives and Archaeology Service. The book is available from Explore the Past, The Hive, Sawmill Walk, The Butts, Worcester, WR1 3PD.

The forward in this book stresses that none of the recipes have been tried or tested, so you do so entirely at your own risk. Nevertheless, they use common ingredients and methods, though you might need to tweak the recipes to get the results you want.

Do you have some stale bread? Use it in a frugal bread pudding.

Boiled Bread Pudding

Reproduced as written in the original recipe, with archive accession code. The photographs of the original documents appear in the booklet.

Ingredients:
   1 1/2 pints of milk
   3/4 pint of bread crumbs
   sugar to taste
   4 eggs
   1oz of butter
   3oz currants
   1/4 teaspoonful grated nutmeg

Make the milk boiling & pour It on the breadcrumbs; let these remain till cold; then add the other ingredients taking care that the eggs are well beaten. Beat the pudding well and put it into a buttered basin. Tie it down tightly with a cloth plunge it into boiling water & boil for 1 1/4 hours – turn it out of basin & serve with sifted sugar. Any odd pieces or scraps of bread answer for this pudding but they sh. be soaked over night & when wanted for use th. have the water well Squeezed from them.

Boiled Bread Pudding: c 20th C. WAAS* 705:884 BA8220/10

From the estate of Edward Lawrence Edwards, fruit grower of Evesham.

What better than to top this with some creamy and spicy very fine custard.

Very Fine Custard

take a quart of creame boyle it with whole mace or what spice you pleas then beat the youlkes of ten eggs and five whites mingle them with a lit tell cream and when your creame is all most cold put your eggs into it and stur them very well then sweet en ih it and take out your custard into a deep dish and bake it then serve it in with french comfits** strewed on it

Very Fine Custard: C.18th C. WAAS* 705.24 BA81/918/250
   From the Berrington family papers
   *Worcestershire Archive and Archaeology Service
   **preserved fruit (try sugared fruit or dried fruit)

PEARS

Spicy Baked Black Pears

Black pears are a type of hard pear called wardens which never fully ripen. Although you wouldn't want to eat them straight off the tree, they have the advantage that they can be stored for months, then cooked to make both tasty sweet dishes or served with meat. They're particularly associated with Worcestershire.

This recipe is reproduced with kind permission from *Worcester-*

shire Orchards[26]. Black pears can be hard to get hold of, but you can use other warden pears instead.

Ingredients:
　6 large firm warden pears
　1/2 - 3/4 pint/300 - 450ml red wine
　1oz/28g brown sugar
　Pinch ground cinnamon, ginger and saffron

Method:

- Peel the pears and place in an oven proof dish
- Mix the red wine with the brown sugar and spices and pour over the pears
- Bake in the oven at 180°C/350°F/Gas Mark 4 until tender (this can be up to 2 hours in the case of black pears)

Baked Red Cabbage and Black Pear in Cider Vinegar

This recipe is reproduced with kind permission of Wade Muggleton from *The Worcester Black Pear*[27].

Ingredients:
　1 medium red cabbage - sliced
　2 white onions – peeled and sliced
　2 Worcester Black pears – peeled and thinly sliced

---

[26] *The Worcester Black Pear* - http://www.worcestershireorchards.co.uk/black-pear

[27] *The Worcester Black Pear* by Wade Muggleton with recipes by Clare Tibbits - https://www.marcherapple.net/product/new-book-the-worcester-black-pear-by-wade-muggleton

120 ml cider vinegar (or try perry vinegar if available!)
3 tablespoons water
(recipe by Clare Tibbits)

Method:

- In a medium-sized casserole cover the bottom with half the sliced red cabbage
- Add the onion, pear and half the vinegar
- Add the remaining red cabbage and the rest of the vinegar
- Cover with a lid and bake in a medium to hot oven (for at least an hour) until tender

About the Author

Liz Pearson Mann writes about being rooted in landscape, traditional culture and evergreen skills. She's an archaeologist who has spent many years producing data on the lives of farmers and the food they've produced from the ground beneath their feet. She's a 'doer' who grows food on an allotment and in the garden at home, knits, spins yarn, and makes her own clothes. She lives in Worcester, in the English West Midlands with her husband and cat.

You can connect with me on:

🌐 https://lizpearsonmann.com
🐦 https://twitter.com/lizpearsonmann
📎 https://www.instagram.com/lizpearsonmann

Subscribe to my newsletter:

✉ https://lizpearsonmann.ck.page/elya-sign-up